Berichte aus dem Institut für Mehrphasenströmungen

Band 8

# Transport Processes at Taylor Bubbles in Vertical Channels

**Vom Promotionsausschuss der**
**Technischen Universität Hamburg**
zur Erlangung des akademischen Grades
Doktor-Ingenieur (Dr.-Ing.)

genehmigte Dissertation

von
Sven Kastens

aus
Bremen

2021

**Bibliografische Information der Deutschen Nationalbibliothek**
Die Deutsche Nationalbibliothek verzeichnet diese Publikation in der Deutschen Nationalbibliografie; detaillierte bibliografische Daten sind im Internet über http://dnb.d-nb.de abrufbar.
1. Aufl. - Göttingen: Cuvillier, 2021
Zugl.: (TU) Hamburg, Univ., Diss., 2021

**Gutachter:**

Prof. Dr.-Ing. Michael Schlüter

Prof. Dr. Andreas Liese

Prof. Dr. Akio Tomiyama

**Vorsitzender des Prüfungsausschusses:**

Prof. Dr. Raimund Horn

**Tag der mündlichen Prüfung:**

04.03.2021


Nonnenstieg 8, 37075 Göttingen
Telefon: 0551-54724-0
Telefax: 0551-54724-21
www.cuvillier.de


1. Auflage, 2021
Gedruckt auf umweltfreundlichem, säurefreiem Papier aus nachhaltiger Forstwirtschaft.

ISBN 978-3-7369-7404-3
eISBN 978-3-7369-6404-4

# Danksagung

Vielen Dank an all die vielen lieben Menschen, die mich während meiner wissenschaftlichen Karriere begleitet und unterstützt haben. Besonders möchte meinen Doktorvater Prof. Dr.-Ing. Michael Schlüter für die Betreuung, das Vertrauen und die Freiheiten danken, durch die ich meine Forschung frei gestalten konnte. Ebenso möchte ich Prof. Dr. Akio Tomiyama für die vielen anregenden Diskussionen ab der ersten Stunde danken. Durch Ihren Rat, Ihre kritischen Fragen und Ihre Herzlichkeit konnte ich meine Forschung und mich selbst weiter entwickeln. Ich danke ganz herzlich Herrn Prof. Dr. Andreas Liese für die Begutachtung meiner Arbeit und Herrn Prof. Dr. Raimund Horn für die Übernahme des Vorsitzes des Prüfungsausschusses.

Vielen lieben Dank an meine Kolleginnen und Kollegen am Institut für Mehrphasenströmung für die Unterstützung und die tolle Zeit vor, während und nach dem Labor. Insbesondere Dr. Jens Timmermann, Prof. Dr. Alexandra von Kameke, Dr. Sophie Rüttinger und Simon Mattes. Meinen Kollegen an der Kobe University danke ich zutiefst für die unendliche Gastfreundschaft und die unermüdlichen Einsatz bei unseren gemeinsamen Experimenten und Publikationen. Vielen Dank Dr. Shogo Hosoda, Dr. Jiro Aoki, Dr. Yohei Hori, Dr. Shohei Sasaki, Prof. Dr. Shigeo Hosokawa und Prof. Dr. Kosuke Hayashi. Meinen Studierenden und studentischen Hilfskräften danke ich für die tatkräfte Unterstützung und hoffe, dass ich ihnen etwas für ihre eigenen Wege mitgeben konnte. Danke Krischan Sandmann, Lisa Christin Watter, Carolin Lohmann, Elettra Sahoo, Caroline Otto, Obinna Kingsley Uwa, Zeynep Percin, Lennard Hasskamp, Finn Leib, Helen Guttowski, Vincent Settler, Catrin Compart, Lars Nelke, Shunya Tanaka, Motoki Iwakiri, Riho Sekimiztu, Yusuke Noguchi.

Insbesondere danke ich meinen Eltern für die Ermutigungen und die bedingungslose Unterstützung während meiner beruflichen Entwicklung. Alex für die konstruktiven Diskussionen und Korrekturschleifen in der finalen Phase. Meinem Bruder und meinen Freunden für die Geduld und für das vorgespielte Interesse an meiner Grundlagenforschung. Vorallem danke ich Nadine für Ihre Geduld während der schlechteren Phasen und die Aufmunterung und Ablenkung durch Lachen, Essen und Reisen. Ohne Dich hätte ich so manche Rückschläge nicht so schnell überstanden.

Danke

Für Nadine und Henry Flynn ...

# Table of contents

# List of figures

# List of tables

# Nomenclature

## Roman Symbols

| | | |
|---|---|---|
| $A$ | $m^2$ | Area |
| $a$ | $m^{-1}$ | Specific interfacial area |
| $C_D$ | - | Drag coefficient |
| $C_i$ | $mol\,m^{-3}$ | Molar concentration of specie i |
| $D$ | m | Channel diameter |
| $d_{32}$ | $m$ | Sauter diameter |
| $d_{eq}$ | $m$ | Sphere volume equivalent diameter |
| $D_L$ | $m^2\,s^{-1}$ | Diffusion coefficient |
| $Eo$ | – | Eötvös number |
| $F_B$ | N | Buoyancy force |
| $F_D$ | N | Drag force |
| $Fr$ | – | Froude number |
| $g$ | $kg\,m\,s^{-2}$ | grav. constant |
| $Ga$ | – | Galilei number |
| $H^*$ | – | Henry coefficient |
| $I_{ex}$ | – | Excitation Intensity |
| $j$ | $kg\,s^{-1}\,m^{-2}$ | Mass flux |

| | | |
|---|---|---|
| $K_f$ | – | Fluid number |
| $k_L$ | $\mathrm{m\,s^{-1}}$ | Mass transfer coefficient |
| $K_{SV}$ | – | Stern-Volmer constant |
| $L$ | m | Characteristic length |
| $l$ | m | Length |
| $M$ | mol | Mol |
| $Mo$ | – | Morten number |
| $n_i$ | – | Refractive index |
| $p$ | Pa | Pressure |
| $Pe$ | – | Peclet number |
| $Q$ | – | Magnification factor |
| $R$ | m | Channel radius |
| $r$ | m | Radius |
| $R_m$ | $\mathrm{J\,mol^{-1}\,K^{-1}}$ | Molar, universal or ideal gas constant |
| $Re$ | – | Reynolds number |
| $s$ | m | Particle displacement |
| $Sc$ | – | Schmidt number |
| $Sh$ | – | Sherwood number |
| $T$ | K | Temperature |
| $t$ | s | Time |
| $u, v, w$ | $\mathrm{m\,s^{-1}}$ | Velocity components in x, y and z direction |
| $V$ | $\mathrm{m^3}$ | Volume |
| $v_B$ | $\mathrm{m\,s^{-1}}$ | Bubble velocity |
| $We$ | – | Weber number |

| | | |
|---|---|---|
| $X_i$ | – | Mole fraction of species i |
| $z^*$ | mm | Local z-axis coordinate |

## Greek Symbols

| | | |
|---|---|---|
| $\delta$ | m | Film thickness |
| $\varepsilon_g$ | – | Gas hold up |
| $\eta$ | $\mathrm{kg\,m^{-1}\,s^{-1}}$ | Dynamic viscosity |
| $\Gamma$ | $\mathrm{mol\,m^{-2}}$ | Interfacial saturation concentration |
| $\kappa$ | $\mathrm{\mu S\,cm^{-2}}$ | Conductivity |
| $\kappa_v$ | – | Viscosity ratio |
| $\lambda$ | – | Diameter ratio |
| $\nu$ | $\mathrm{m^2\,s^{-1}}$ | Kinematic viscosity |
| $\phi$ | $^\circ$ | Angle |
| $\psi$ | $^\circ$ | Angle |
| $\rho$ | $\mathrm{kg\,m^{-3}}$ | Density |
| $\sigma$ | $\mathrm{N\,m^{-1}}$ | Interfacial tension |
| $\tau$ | s | Contact time |

## Subscripts

| | |
|---|---|
| $\infty$ | Free rising |
| abs. | Absolute |
| atm. | Atmospheric |

| | |
|---|---|
| B | Bubble |
| c | Continuous phase |
| crit. | Critical |
| d | Dispersed phase |
| em. | Emitting |
| eq. | Equivalent |
| ex. | Excitation |
| film | Fluid film |
| g | Gaseous phase |
| h | Hydraulic |
| i | Species i |
| in. | Injected |
| l | Liquid phase |
| loc. | Local |
| P | Particle |
| Q | Quenched |
| rel. | Relative |
| sat. | Saturation |
| SR | Surface renewal |
| tot. | Total |
| V | Molar |
| wake | Wake |

**Acronyms / Abbreviations**

| | |
|---|---|
| CCD | Charged coupled device |
| CCF | Counter current flow |
| CFD | Computational Fluid Dynamics |
| CMOS | Complementary Metal Oxide Semiconducto |
| $CO_2$ | Carbon dioxide |
| DMSO | Dimethyl Sulfoxide |
| fps | Frames per second |
| LED | Laser emitting diode |
| LIF | Laser Induced Fluorescence |
| $N_2$ | Nitrogen |
| Nd:YAG | Neodym-doped Yttrium Alumina Granate |
| Nd:YLF | Neodym-doped Yttrium-Lithium-Fluorid |
| $O_2$ | Oxygen |
| pH | pH value |
| PIV | Particle Image Velocimetry |
| Px | Pixels |
| SFV | Spatial Filter Velocimetry |

# Abstract

Gas / liquid contact apparatuses are widely used in chemical, biochemical or pharmaceutical industry to provide and transfer species as reactant from the gaseous phase to pre-dissolved reactants in the liquid phase enabling a preferred reaction. The selectivity and yield of chemical reactions depend on the timescales of the mass transfer of this gas specie, e.g. $O_2$ or $CO_2$, across the interface, the mixing inside the liquid with the bulk phase and the reaction kinetics. For processes with fast chemical reactions, this mass transfer can be a limiting factor, where the reactant supply from the gaseous phase is leaking and needs to be enhanced. The global and local transport are complex interlinked processes and therefore in practice in reactor design industry, mostly empirically correlated. For a better understanding, a secure control of the overall process and a more efficient reactor design, the local transport processes caused by fluid dynamics and the mass transfer at gas / liquid interfaces need to be investigated. Additionally, due to the application of process additives or occurring impurities in substrates, there are surface active agents, which affect important fluid system parameters like surface tension or local boundary conditions by forming molecular layers at the interface. In the past, there have been many investigations concerning the rise behavior, shape deformation, break-up and foaming of bubbly flows with surface active agents, but entangling the influence on the local transport processes is still challenging for detailed experimental investigations.

In this work the global and local fluid dynamics and mass transfer are investigated in clean and contaminated systems to provide a deeper knowledge and understanding of each process. For the detailed investigations Taylor bubbles in vertical channels are used because they overcome the problem of dynamic shape deformation and complex 3D rise trajectories. Additionally, Taylor bubbles have a volume independent rise velocity and can therefore be kept in counter current flow easily, which make them an ideal multiphase flow system for detailed optical investigations with a high temporal and spatial resolution even for long-term measurements. The measuring techniques applied in this work are high-speed Shadowgraphy, 2D2C Particle Image Velocimetry and p-2D Laser Induced Fluorescence to observe the rise and shrinking behavior of the bubbles, the local velocity fields at the interface and the local concentration fields behind the bubbles and at their interfaces. The combination of local and global techniques can therefore be used to validate the results in e.g. mass balances and therefore to validate the measuring techniques.

The main findings of this work are the characterization of small, intermediate and Taylor bubbles in the $D = 7$ mm channel by the bubbles' local maximal radius in the bottom region and the corresponding local minimal fluid film thickness. The local bubble geometry is the parameter with the major effect on the bubble rise velocity, even dominating the bubble volume. This generates an untypical slope of the rise velocity over bubble equivalent diameter, where the slope turning points define the boarders of the named bubble types and proofs the volume independent rise velocity of Taylor bubbles. Additionally, a Sherwood $Sh_D$ correlation is derived that describes the mass transfer of $CO_2$ Taylor bubbles over a wide range of channel diameters $D = 5.5 - 25$ mm, where $D$ is established as characteristic length. The mass transfer from single $CO_2$ Taylor bubbles in vertical mini-channels was measured for various channel hydraulic diameters $D_h$. Sherwood numbers, $Sh_D$, using $D_h$ as characteristic length, are well-correlated in terms of the Eötvös number $Eo$. The proposed $Sh_D$ correlation can predict the long-term dissolution process of Taylor bubbles, while taking into account counter diffusion of pre-dissolved gases like $O_2$ & $N_2$.
The analytics of the local fluid dynamics via high-speed 2D2C-PIV enable the identification of significant different wakes structures, like laminar, toroidal and turbulent mixing wake structures behind Taylor bubbles in channels of diameters of $D = 6, 7$ and 8 mm. Those wake structures depend on the Reynolds number and can be transferred to solvents like acetonitrile or methanol when keeping $Eo$ constant.

Another main finding, is the time evolution of dissolved $CO_2$ concentration in the wake structure behind a Taylor bubble measured via p-2D-LIF during a long-term dissolution process. The local concentration is reduced by the mole fraction shift from pure $CO_2$ to gas mixtures due to counter diffusion of $N_2$ & $O_2$, which directly effects the mass transfer according to the Two Film Theory. Additionally, a mass balance was established between the globally measured bubble volume reduction and the integral of the local measured $CO_2$ concentration field, which validates the high accuracy of the applied measuring techniques and provides an experimental method for the estimation of local reactant concentration in multiphase systems.

On the basis of this work, it was possible to establish the Taylor bubble set up as a guidance tool for chemical systems within the priority program SPP1740 "Reactive bubble flows" of the German Research Foundation. In future, reactive multiphase systems could be characterized within this complexity reduced experiments to get first insights of the relation between reaction and mixing timescales. Additionally, this tool minimizes the risk of hazardous accidents in chemical industry, because it requires minimal volumes of reactants and solvents. The extension of reactive bubbly flows to high pressure systems will require the shift from translucent materials like glass channels to steel experimental apparatus. Therefore, local and global transport processes at Taylor bubbles need to be validated with state of art tomography techniques for a new chapter of research on industrial relevant reactive multiphase flows.

# Zusammenfassung

Gas / Flüssigkeits-Kontaktapparate werden in der chemischen, biochemischen oder pharmazeutischen Industrie häufig verwendet, um Reaktanten aus einer Gasphase in eine Flüssigphase zu übertragen, um dort mit anderen vorgelösten Reaktanten zu einem gewünschten Produkt zu reagieren. Die Selektivität und Ausbeute von chemischen Reaktionen hängen von den Zeitskalen des Stoffübergangs dieser Gasspezien, z.B. $O_2$ oder $CO_2$ über die Phasengrenzfläche, der Vermischung innerhalb der Flüssigkeit mit der Bulk-Phase und der Reaktionskinetik ab. Bei Prozessen mit schnellen chemischen Reaktionen kann der Stoffübergang der limitierende Faktor für den gesamten Prozess sein, bei dem die Reaktantenzufuhr aus der Gasphase unzureichend ist und verbessert werden muss. Der globale und der lokale Stofftransport sind komplexe miteinander verbundene Prozesse und daher in der Praxis der Reaktordesignindustrie meist empirisch korreliert. Zum besseren Verständnis der Prozesse, einer sicheren Prozessskontrolle und eines effizienteren Reaktordesigns, müssen die lokalen Transportprozesse, die durch die Fluiddynamik und den Stofftransport an der Phasengrenzfläche zwischen Gas und Flüssigkeit untersucht werden. Zusätzlich gibt es aufgrund von Prozessadditiven oder auftretenden Unreinheiten der Substrate oft oberflächenaktive Substanzen, die molekulare Schichten an der Grenzfläche bilden und damit Parameter des Fluidsystems, wie Oberflächenspannung oder lokale Randbedingungen beeinflussen. In der Vergangenheit gab es viele Untersuchungen zum Aufstiegsverhalten, zur dynamischen Blasendeformation, zum Aufbrechen und Koaleszieren von Blasenströmungen mit oberflächenaktiven Substanzen, aber die gegenseitige Beeinflussung der lokalen Transportprozesse ist für detaillierte experimentelle Untersuchungen immer noch eine Herausforderung.

In dieser Arbeit werden die globale und lokale Fluiddynamik und der Stofftransport in sauberen und kontaminierten Systemen untersucht, um ein besseres Verständnis für die Interaktion der Transportprozesse zu ermöglichen. Mittels detaillierten Messungen an Taylorblasen in vertikalen Kanälen können die Transportprozess an der Phasengrenzfläche ohne die Probleme der dynamischen Blasendeformation und komplexer 3D-Aufstiegstrajektorien untersucht werden. Zusätzlich haben Taylorblasen eine volumenunabhängige Aufstiegsgeschwindigkeit und können daher im Flüssigkeitsgegenstrom örtliche fixiert werden, was sie zu einem idealen Mehrphasenströmungssystem für detaillierte optische Untersuchungen mit einer hohen zeitlichen und räumlichen Auflösung macht; und dies sogar für ungewöhlich lange Messzeiten. Die in dieser Arbeit angewendeten Messtechniken sind High-Speed Shadowgraphy, 2D2C-Particle-Image-Velocimetry und planare-2D Laser Induced Fluorescence, um das Aufstiegs- und Schrumpfungsverhalten der Blasen, die lokalen Geschwindigkeitsfelder und die lokalen Konzentrationsfelder der gelösten Gasspezie an der Phasengrenzfläche und bei der Vermischung im Nachlaufgebiet der Blasen zu untersuchen. Die Kombination von lokalen und globalen Messtechniken ermöglicht daher die Einzelergebnisse und Messtechniken mittels einer Massenbilanzen zu validieren.

Die wichtigsten Ergebnisse dieser Arbeit sind die Charakterisierung kleiner, mittleren und Taylorblasen im $D = 7$ mm Kanal mittels deren Aufstiegsgeschwindigkeit, welche sich durch den lokalen maximalen Blasenradius im unteren Bereich der Blasen und der entsprechenden lokalen minimalen Flüssigkeitsfilmdicken begründen lässt. Die lokale Blasengeometrie ist der Parameter mit dem größten Einfluss auf die Blasenanstiegsgeschwindigkeit und dominiert sogar das Blasenvolumen und damit die Aufstiegskraft. Dies erzeugt eine untypische Verlauf der Aufstiegsgeschwindigkeit über dem äquivalenten Blasendurchmesser, wobei die Extremwerte die Grenzen der genannten Blasentypen definieren und auch die volumenunabhängige Aufstiegsgeschwindigkeit von Taylorblasen erklären. Zusätzlich wird eine Sherwood $Sh_D$ -Korrelation etabliert, die den Stoffübergang von $CO_2$ Taylorblasen über einen weiten Bereich von Kanaldurchmessern $D = 5,5 - 25$ mm beschreibt. Der Stoffübergang von einzelnen $CO_2$ Taylorblasen in vertikalen Kanälen wurde für verschiedene hydraulische Kanaldurchmesser $D_h$ gemessen. Die Sherwood-Zahl $Sh_D$ ist abhängig von der Eötvös-Zahl $Eo$ der Taylorblasen, wenn beide mit charakteristische Länge $D_h$ definiert sind. Die erstellte $Sh_D$ -Korrelation kann den Auflösungsprozess von Taylorblasen vorhersagen, während die Gegendiffusion von vorgelösten Gasen wie $O_2$ & $N_2$ berücksichtigt wird.
Die Analyse der lokalen Fluiddynamik über High-Speed 2D2C-PIV ermöglicht die Identifizierung von signifikant unterschiedlicher Nachlaufstrukturen, wie laminarer, toroidaler und turbulenter Nachlaufstrukturen hinter Taylorblasen in Kanälen mit Durchmessern von $D = 6,7$ und 8 mm. Diese Nachlaufstrukturen und die damit verbundenen Zeitskalen hängen von der mit der Aufstiegsgeschwindigkeit gebildeten Reynolds-Zahl ab und können auf relevante Lösungsmittel der chemischen Industrie, wie Acetonitril oder Methanol, durch konstanthalten der $Eo$-Zahl übertragen werden. Ein weiteres wichtiges Ergebnis ist die zeitliche Entwicklung der Konzentrationsfelder von gelöstem $CO_2$ in der Nachlaufstruktur hinter Taylorblasen, welche über p-2D-LIF während langzeitlichen Auflösungsprozessen

gemessen werden kann. Die $CO_2$-Konzentration in der Blase wird durch die Verschiebung der Molenbrüche, aufgrund der Gegendiffusion von $N_2$, $O_2$, von reinem $CO_2$ zu Gasgemisch verringert. Hierdurch wird der Stoffübergang gemäß der Zwei-Film-Theorie direkt beeinflusst und die Konzentration des gelösten $CO_2$ in der Flüssigkeit an der Phasengrenzfläche reduziert sich mit der Zeit. Zusätzlich wurde eine Massenbilanz zwischen der global gemessenen Blasenvolumenreduktion und dem Integral des lokal gemessenen $CO_2$ -Konzentrationsfeldes bestätigt, was die hohe Genauigkeit der angewandten Messtechniken unterstreicht. Diese experimentelle Methode kann somit auch zur Abschätzung der lokalen Reaktantenkonzentration in reaktiven Mehrphasensystemen eingesetzt werden.

Auf der Grundlage dieser Arbeit wurde die Taylorblase als Leitexperiment für chemische Systeme im Priority Program SPP1740 "Reaktive Blasenströmungen" der Deutschen Forschungsgemeinschaft etabliert. In Zukunft können reaktive Mehrphasensysteme innerhalb dieser komplexitätsreduzierten Experimente charakterisiert werden, um erste Einblicke in die Zusammenhänge zwischen Reaktions- und Mischzeitskalen zu erhalten. Darüber hinaus minimiert dieses Tool das Risiko von Unfälle in der chemischen Industrie, da nur minimale Mengen an Reaktanten und Lösungsmitteln erforderlich sind, um erste Untersuchungen und Erkenntnisse zu den prozessrelevanten Zeitskalen zu ermöglichen. Die Ausdehnung der Forschung auf reaktive Blasenströme in Hochdrucksystemen erfordert die Umstellung von transparenten Materialien wie Glaskanälen auf druckfeste Edelstahlkanäle in zukünftigen Versuchsanlagen. Daher sollten lokale und globale Transportprozesse an Taylorblasen mit modernsten Tomographietechniken validiert werden, um ein neues Forschungskapitel zu industrierelevanten reaktiven Mehrphasenströmungen voranzutreiben.

# Chapter 1

# Introduction

In many multiphase contact apparatuses in chemical engineering, bio-chemical and pharmaceutical industry, new insights into mass transfer processes across fluidic interfaces are essential for process optimization and reliable process control. The need for high quality drugs in pharmaceutical industry or for sustainable production of specialty and bulk chemicals provides the driving force for more detailed investigations. Multiphase contact apparatuses are used to transfer gaseous reactants from a highly dispersed gas phase across the gas-liquid interface into a liquid solvent. The gas transfer should be immediately and ideal everywhere, since the dissolved gas is going to be consumed by a chemical or biochemical reaction and side reactions or a malnutrition of cells are often minimized by avoiding high gradients of concentration. This means, that the mass transfer performance from the dispersed gaseous into the continuous liquid phase can be the bottle neck for a certain process and therefore be dominant for a whole production and its process efficiency. In general, mass transfer across interfaces can be influenced by global and local boundary conditions. For example, the physical properties of the fluids or global boundary conditions e.g. temperature and pressure can effect the potential of mass transfer, which means the saturation of a fluid by a certain specie, which directly affects the potential mass difference, which can occur at the interface. Also, local boundary conditions can be affected e.g. by fluid dynamic conditions as, for example, kinetic energy of the fluid elements or dissipation and mixing. The transferred mass at the interface into the continuous phase needs to be transferred away from the interface and mixed into the bulk, where it can consumed by a preferred chemical reaction. This already shows, that there are several ways to influence the mass transfer performance by controlling the boundary and local conditions in gas-liquid contact apparatuses.

The coupling of fluid dynamics, mass transfer and chemical reaction have been investigated for decades, but due to its interlinked complexity usually simplified or isolated approaches regarding one or two transfer processes are taken into account only. For example, mass transfer processes across gas-liquid interfaces have been investigated to understand the coupling of hydrodynamics and mass transport processes and to describe and correlate them for certain geometries of gas-liquid flow apparatuses and usually small ranges of process parameters. An important parameter for instance is the bubble rise velocity $v_B$ that affects the fluid dynamic conditions for the dissolution process in bubbly flows. Therefore, for example, free rising bubbles in pure and contaminated systems [Cli78] [Alv05] have been investigated to understand the influence of bubble size and fluid properties such as viscosity and surface tension. Interaction with geometrical reactor parameters has also a strong influence on the flow structure around bubbles, their shapes and behavior [Kri99] [Bro65]. Elongated bubbles in channels are called Taylor bubbles [Tay61] and can be found in multiphase reactors of various scales such as monolith reactors, pipe reactors many more. For a hydrodynamic description of Taylor bubble flows, it is required to distinguish two flow conditions; the rising Taylor bubble which induces a flow field around the bubble by the buoyancy force, and the pressure-driven flow where bubbles are transported through a channel by the flowing liquid. Taylor bubbles are rising by buoyancy as long as the pipe or channel diameter $D$ is larger than the critical inner diameter $D_{crit}$ calculated by the critical Eötvös number $Eo_{Dcrit}$. The Eötvös number

$$Eo_D = \frac{(\rho_l - \rho_g) \cdot g \cdot D_h^2}{\sigma} \tag{1.1}$$

is defined as the ratio of the buoyancy force to the interfacial tension, where $\sigma$ is the interfacial tension, $\rho_l$ and $\rho_g$ are the densities of the liquid and gaseous phases, $g$ is the magnitude of the gravitational acceleration and $D_h$ is the hydraulic diameter of a channel. $D_{crit}$ is about 5.4 mm in a pure air-water system, but $D_{crit}$ is of course strongly depending on the systems surface tension. Transport processes at sequential gas-liquid Taylor flows in small channels ($D_h < D_{crit.}$) are often used in "Lab-On-a-Chip" micro fluidic devices or other simple continuous small scale applications, where pressure differences are needed to provide these flow structures, but those are off-topic in this work. For buoyancy driven flows in channels $D_h > D_{crit.}$, it is known that the rising velocity $v_B$ of Taylor bubbles is independent of the bubble volume. The independence of $v_B$ on the bubble volume enables the suspension of Taylor bubbles at a fixed elevation in vertical channels by an equivalent counter current flow. Filla [Fil81] has shown the applicability of bubble fixation by a counter current flow and measured the liquid side mass transfer of single $CO_2$ Taylor bubbles by a photographic technique in a vertical tapered channel. Esteves and de Carvalho [Sen93] conducted experiments on the mass transfer from very long single Taylor bubbles, of which the ratio of bubble length to $D_h$ is larger than 10, in large channels of $D_h = 32$ mm. Hosoda et al. [Hos14a] carried out experiments on the mass transfer from single $CO_2$ bubbles in vertical pipes of 12.5 - 25mm. They have developed a following Sherwood number correlation for Taylor bubbles rising in vertical pipes according to

$$Sh = \frac{k_L \cdot d_{eq}}{D_L} = (0.49\lambda^2 - 0.69\lambda + 2.1)\, Pe^{1/2} \tag{1.2}$$

where $k_L$ is the mass transfer coefficient, $D_L$ the diffusion coefficient and $\lambda$ the diameter ratio, which is the ratio of the sphere volume equivalent bubble diameter $d_{eq}$ to $D_h$. Here the Peclet number $Pe$ is defined by

$$Pe = \frac{v_B \cdot d_{eq}}{D_L}. \tag{1.3}$$

Eq. 1.2 is valid for $\lambda = 0.6$ and applicable to the prediction of long-term dissolution processes of single $CO_2$ Taylor bubbles in water ($D = 12.5 - 25$ mm).
Though several studies on the mass transfer from Taylor bubbles in channels were carried out as discussed above, our knowledge on mass transfer from single buoyancy-driven Taylor bubbles in smaller channel diameters and on effects of channel cross-sections on the mass transfer are still rudimentary, especially for $D_{crit.} < D_h < 12.5$ mm. Additionally, Taylor bubbles in these channels sizes combine some main advantages compared to free rising bubbles, like the volume independent rise velocity and the suppressed shape dynamics even for comparable large bubbles. Therefore, these Taylor bubbles with low $Eo$ number are the ideal multiphase system to investigate global and local transport processes at rising or fluid dynamically fixed bubbles. This enables even close-ups of the interface during a untypical long time period for bubbly flows.

During past decades, empirical correlations on the characterization of global transport processes and their general interaction in reactive bubbly flows and enabled to design effective industrial processes. Nowadays, the fast development of Computational Fluid Dynamics increased the pressure on the experimental research to provide even local data of transport processes and chemical reactions for validation and to understand the complex interaction of timescales of transport, mixing and reaction for an optimized production with high yield and selectivity. The combination of global and local measuring techniques in the complexity reduced multiphase system of Taylor bubbles enables new insights into fluid dynamics and mass transfer process as well as the overall mass balance of the transferred specie. The transfer from aqueous systems to industrial relevant solvent systems is the key use-case for new, fast and safe methods for reactive process design. The validation of global and local data from CFD and experimental investigations at Taylor bubbles will enable new insights into the interlinked transfer processes of reactive systems and will lead to reliable process control for "transparent" chemical plants in times of industry 4.0.

# Chapter 2

# State of the art and basic knowledge

## 2.1 Gas-liquid flows in channels

In chemical engineering contact apparatuses like bubble columns or pipe reactors (e.g. heat exchangers) are used, for the dispersion of one phase into another, to create a large interfacial area between those phases, enabling a high specie transfer and good mixing conditions. Here flow maps help to choose the ideal flow condition to design a reactor for a given reaction. Usually, the chemical reaction is expected in the liquid phase, where two or more reactants are brought into contact, which were either transported from gas to liquid phase or have been pre-dissolved. Thereby, the mass transfer from the gas into the liquid phase is often a critical point in terms of efficiency but a controllable process by adapting flow rates, interfacial area and mixing [Hew69, Kra17].

If a gaseous phase is dispersed into a liquid, bubbles are generated due to the interfacial tension between the two phases. Here the liquid is defined as continuous phase and the gaseous phase is the dispersed phase. The multiphase regimes can be adapted by the variation of the volume flow rates of both phases, where in gas-liquid flows usually the gas hold up $\varepsilon_g$ is used, which is the ratio of the gas volume $V_g$ to the total volume $V_{total}$ of the system. The occurring regimes are characterized by visual differences and flow phenomena. The flow regimes depend also strongly on the reactor geometry and its solid walls, e.g. vertical channels, solid internals, stirrers or heat exchangers, which define the absolute boundaries of the fluid dynamics system. Typical flow regimes of gas-liquid systems in vertical channels can be seen in Fig. 2.1, which are described as bubbly, film/annular, plug or slug flow. The actual visual display of the regimes can vary, depending on fluid viscosity and interfacial effects, which become actually more dominant, with smaller reactor geometry. To enable local investigations in multiphase sytems, a reduction of the complexity is needed to reduce stochastic phenomena, to establish quasi steady conditions and to ensure a high experimental

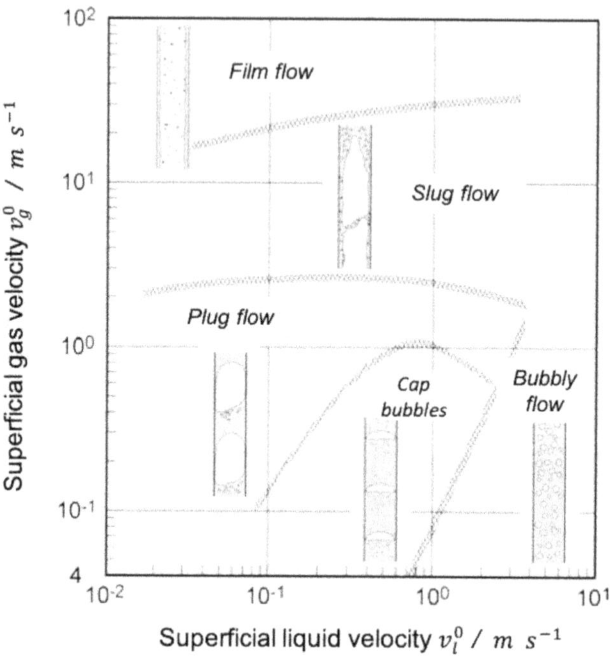


Fig. 2.1 Multiphase flow regimes in vertical pipes in terms of superficial gas velocity $v_g^0$ vs. superficial liquid velocity $v_l^0$ after Kraume [Kra17].

reproducibility. These are required conditions to enlarge the possibility of fundamental research to identify general interactions and dependencies of local transport phenomena.
In the following chapter, the fundamental processes of fluid dynamics and mass transfer are described on simplified single bubble models, before the influence of channel walls on the transport processes is introduced. Afterwards, the main scientific findings of historical and recent investigations on rising bubbles in vertical channels will be introduced, before the lack of knowledge in this research field leads to the scientific goals of this work.

## 2.2 Transport processes in bubbly blows

### 2.2.1 Fluid dynamics at free rising bubbles

Industrial multiphase applications usually require a high gas hold up $\varepsilon_g$, which is the ratio between the gas volume $V_g$ and the total reactor volume $V_{tot.}$ and a strong mixing to ensure a high specific interfacial area $a$ enabling a high mass transfer from gas to liquid phase. The high phase interaction and fluid mixing leads to a high number of homogeneous distributed small bubbles in those reactors and therefore establish homogenous conditions for reactant transfer and following reactions. But those bubbles cannot be approximated as a single free rising bubbles. Bubbles in swarms or plumes influence massively their movement, shape dynamics and wake structures and therefore the mixing inside the liquid bulk phase. These combined effects are influencing the local temporal mass transfer performance, the selectivity of chemical reactions and therefore the overall process efficiency. The empirical correlations for a process design are usually depending on the simplified rise regime of single bubbles, because those are much easier to reproduce and to investigate with e.g. optical techniques than bubble swarms with their stochastic behavior. The investigation of local phenomena of transport processes in complex interlinked stochastic multiphase systems would require an enormous experimental or numerical effort. To ensure the accessibility of local mass transfer processes or even chemical reactions, usually strongly simplified multiphase systems are chosen. In this way the experimental reproducibility is enhanced, which aids to verify observed phenomena and derive the cause and impact of the interrelation of transfer processes. Therefore, fundamental research in multiphase flows is usually done with single bubble experiments to understand the impact of parameter variations on specific processes, before the findings are transferred to more complex gas-liquid flows with a higher grade of interaction and stochastic behavior.

The movement or rising of gaseous particles or bubbles in multiphase gas-liquid systems has been investigated since decades, by many scientists describing the fundamentals for various systems and conditions e.g. Brauer [Bra71] and Clift [Cli78]. This work focuses on water based systems with different gaseous phases under standard conditions ($T$, $p$, etc.) and excludes special phenomena of viscous effects, like non-newtonian fluids. In the following, the main fundamental findings are summarized and introduced for additional details the reader is refered to basic literature from e.g. Clift et al. or Brauer.

In general, particle movement in multiphase systems is described as the movement of a relatively small separated phase volume, which is dispersed into another continuous phase, where the particle can be solid (excluded here) or fluid (either gas or liquid) and the continuous phase is fluid (either gas or liquid). Depending on the density $\rho_i$ of both phases, the particle can sink or rise in the continuous phase. This is described with the buoyancy force

$$F_{B,i} = g \cdot \Delta\rho_i \cdot V_i \tag{2.1}$$

where, $i$ is the phase indicator, $\rho$ is the density and $V$ is the volume of the phase. If a particle is dispersed inside a fluid and has a lower density than the continuous fluid it displaces, the force balance will be negative and therefore pointing against the gravitational acceleration $g$, which leads the particle to rise. If the density of both phases is the similar or identical, the particle is neutrally buoyant and follows the motion of the continuous phase. The movement of a spherical particle that has a relative velocity to the surrounding liquid velocity, e.g. a small bubble, is often parameterized using the dimensionless Reynolds number

$$Re = v_P \cdot d_B / \nu \tag{2.2}$$

which is determined as the ratio of inertial to viscous forces. Here, $v_p$ is relative velocity of the bubble with a diameter $d_B$ and $\nu$ is the viscosity of the continuous phase. In the following, the moving dispersed fluid particle, indicated by the subscript $P$ is generally a gaseous bubble, indicated with the subscript $B$, and the continuous phase is an aqueous solution.

The movement of a bubble can be described by the force balance of the buoyancy force Eq. 2.1, here for a spherical bubble,

$$F_B = g\,\Delta\rho\, d_B^3 \,\frac{\pi}{6} \tag{2.3}$$

and the counter acting drag force,

$$F_D = C_D\, d_B^2\, \frac{\pi}{4}\, \frac{\rho_l\, v_B^2}{2} \tag{2.4}$$

which is mainly effected by the fluid properties like density $\rho_l$, the bubble diameter $d_B$, the drag coefficient $C_D$ and the relative bubble velocity $v_{B,rel}$. With the force balance between Eq. 2.3 and Eq. 2.4 the rise velocity of a bubble can be calculated as

$$v_B = \sqrt{\frac{4}{3}\frac{d_B\, g\, \Delta\rho_i}{C_D\, \rho_l}}. \tag{2.5}$$

In stagnant liquid $v_l$, the absolute rising velocity $v_{B,abs}$ of a bubble is equal to the relative velocity $v_{B,rel}$. For flowing liquids $v_l \neq 0$ the relative motion has to be taken into account according to

$$v_{B,rel} = v_{B,abs} - v_l. \tag{2.6}$$

The influence of the physical properties of the gaseous phase on the bubble movement, such as density $\rho_g$ and viscosity $\eta_g$, can usually be neglected compared to the properties of the fluid such as density $\rho_l$, surface tension $\sigma$ and dynamic viscosity $\eta_l$. The drag of any particle depends on friction between the dispersed particle and the continuous phase. For low Reynolds numbers the friction dominates and fluid particles have the same drag coefficient than rigid particles according to Stokes law, where

$$C_D = \frac{24}{Re}; \; Re \leq 0.1 \tag{2.7}$$

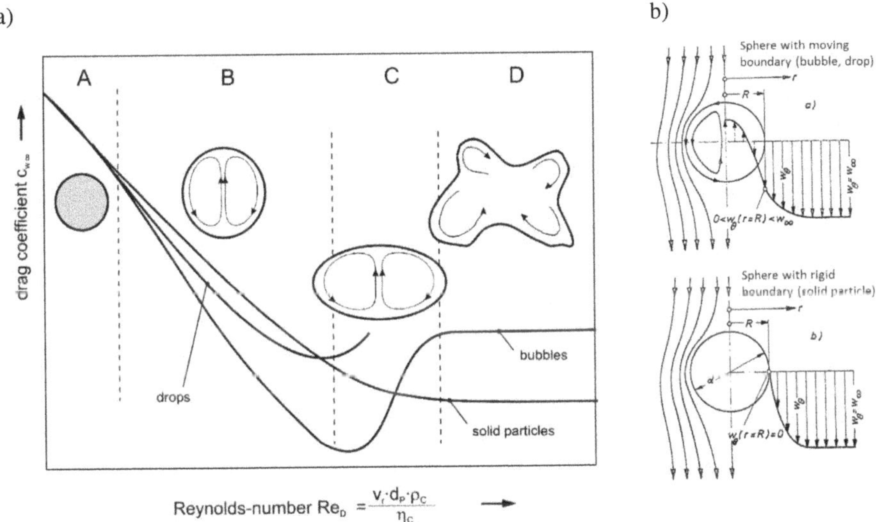


Fig. 2.2 Interface mobility and fluid properties affecting the drag coefficient of particles. a) Drag-coefficient of solid (no-slip), fluid and gaseous particles [VDI15] ; after Peebles [Pee53]; b) Theoretical comparison of the interface mobility of fluid and solid particles and their effect on the local velocity field [Bra71].

gas, liquid and solid particles have the same $C_D$ (compare regime A in Fig. 2.2 a) ). In regime B, fluid particles start to develop internal circulation flows, induced by friction between the phases at the interface, which changes the boundary condition from no-slip to slip at the interface in Fig. 2.2 b). Therefore, the higher liquid velocity at the interface leads to a reduced friction and to a lower $C_D$ than for solid particles. The overall $C_D$ for particles in regime B is given as

$$C_{D,tot} = \frac{8}{Re}\left(\frac{2+3\kappa}{1+\kappa}\right) \tag{2.8}$$

the dynamic viscosity of dispersed ($\eta_d$) and continuous phase($\eta_c$), where its ratio is named $\kappa$

$$\kappa = \frac{\eta_d}{\eta_c}. \tag{2.9}$$

With increasing *Re* number of fluid particles, those are first being deformed to ellipsoidal and further to irregular shapes, which influences the particle cross-sectional area in flow direction and leads to dynamic changes in Drag-coefficients compared to solid particles with constant cross-sectional areas [Pee53], as shown in Fig. 2.2 a). According to Fig. 2.2) regime B, it is obvious that $C_D$ of spherical droplets lay in between solid particles ($\eta_p \rightarrow \infty \Rightarrow \kappa \rightarrow \infty \Rightarrow C_{D,p} = \frac{8}{Re} \cdot 3 = \frac{24}{Re}$) and bubbles ($\eta_b \rightarrow 0 \Rightarrow \kappa \rightarrow 0 \Rightarrow C_{D,p} = \frac{8}{Re} \cdot 2 = \frac{16}{Re}$). The ability of a fluid particle to deform from a sphere into an ellipsoid or irregular shape is affected by the fluid properties and the interfacial tension of the fluids. Usually, there are three dimensionless numbers in fluid mechanics to describe the movement and shape of dispersed fluid particles; the earlier defined Reynolds number, the Eötvös number 1.1 and the Morton number. The ratio of viscous forces and surface tension $\sigma$ defines in the Morton number

$$Mo = \frac{g\, \eta_c^4\, \Delta\rho}{\rho_c^2\, \sigma^3} = \frac{We^2}{Re^3 \cdot Fr}, \tag{2.10}$$

which depends only on fluid properties. The Morton number can also be expressed by the ratio of other dimensionless number like the Weber number *We*

$$We = \frac{\rho_c\, v_B^2\, d_B}{\sigma}, \tag{2.11}$$

and the Froude number

$$Fr = \frac{v}{\sqrt{g\, d_B}}. \tag{2.12}$$

But not only the fluid properties itself can affect the behavior of dispersed bubbles and drops, also impurities or process additives with surface active parts can reduce the surface tension or the interfacial mobility, which leads to different rise velocities for bubbles in pure and contaminated water. This effect is shown in Fig. 2.3 and can additionally change the transition of the rise regime B to C and C to D for free rising bubbles (compare Fig. 2.2). The reason for this effect of contamination are introduced in the following chapter.

a)

b)

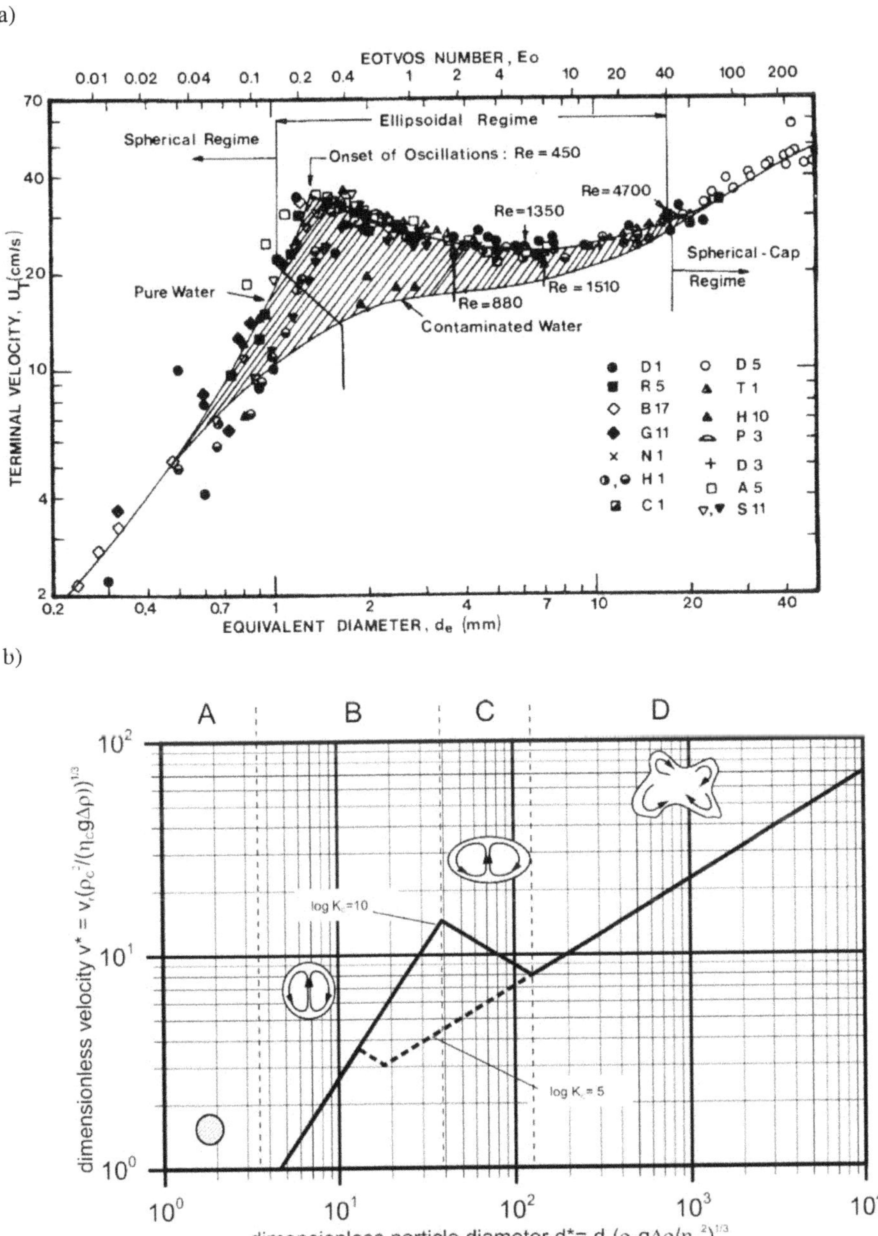


Fig. 2.3 Bubble rise velocity depending on their diameter, shape and fluid number $K_f$ or contamination. a) Bubble rise regimes in clean and contaminated water [Cli78]; b) Bubble regimes depending of diameter, shape and fluid number (dimensionless) [VDI15]

**Effects of surfactants on fluid dynamics**

Surface active agents are molecules with hydrophobic and hydrophilic parts and therefore prefer to adsorb and distribute over the interfaces of aqueous gas-liquid systems, like seen in Fig. 2.4. Those surfactants can be either unwanted impurities of the system or even wanted process additives and generally affect interfacial tension and interface mobility. Therefore, to guarantee reliable process control, it is required to check the effects of surfactants on the transport processes in advance, to clarify if they can be neglected or how their concentration influences fluid properties. Adsorbed surfactants generally reduce the surface tension of interfaces, which make them easier to be deformed and therefore influence the local conditions and transport processes at fluid interfaces. Adsorbed molecules can be moved and accumulated by shear flows, acting on the interface. On the other hand, these molecules try to reach a lower state of energy, where the area density would be uniform. The resulting force, by the area density differences of the adsorbed molecules, is known as Marangoni force. Depending on the interfacial density of surfactant molecules, their effect varies e.g. the interface mobility can be rigid for very high surfactants concentration close to saturation, but usually lower states occur due to the balancing of local forces. In the following some experimental and numerical investigations on qualitative and quantitative effects of surfactants on the fluid dynamics of bubbles and droplets will be introduced.

The surfactant molecules that hit the interface at the bubble front, are sheared to the bubble rear, where they accumulate and may desorb again, if the local interfacial saturation is reached. This model of surfactants accumulation is also known as "stagnant" or "rigid cap model" like seen in Fig. 2.4 from Dukhin et. al [Duk16], where $\Psi$ is the angle of the "stagnant" molecule layer at the bubble rear.

These layers can have significant barrier effects to transport processes and a first study

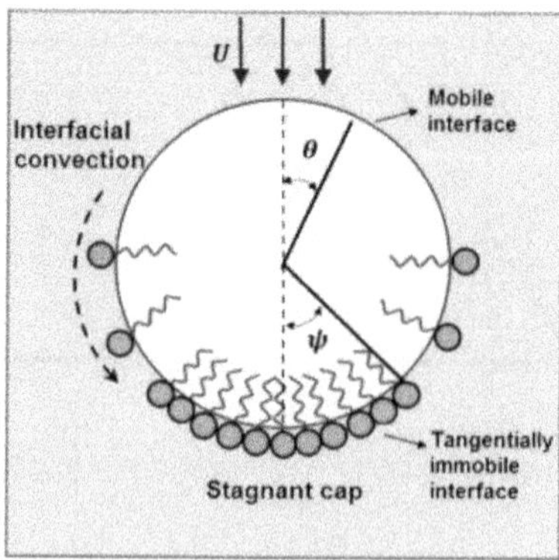


Fig. 2.4 Stagnant cap model of accumulating surfactants on rising bubbles [Duk16]

has been published by Lamer and Healy in Science in the mid 60s [LM65]. They found a changed interfacial viscosity, where the mobility of the interface was reduced (compare Fig. 2.2 a) & b) ), which leads to smaller rise velocities of bubbles especially in the dimension of $d_p = 1$ to 5 mm (compare contaminated line in Fig. 2.3 a)).

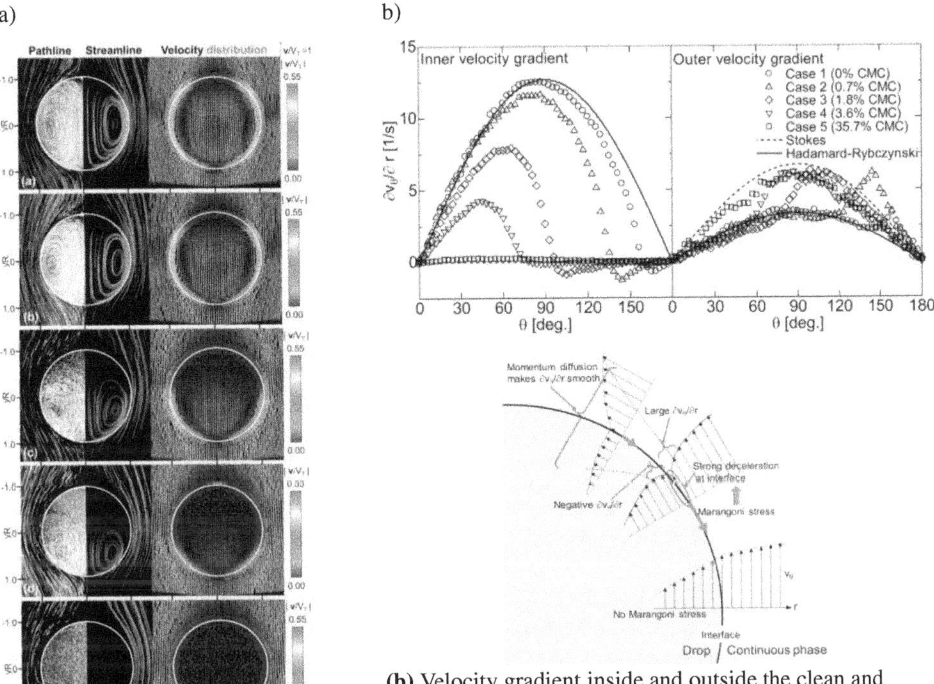


**(a)** Effect of surfactants on the internal circulation and slip condition of falling droplets.

**(b)** Velocity gradient inside and outside the clean and contaminated falling droplets & the schematic Marangoni stress model

Fig. 2.5 Surfactants reduce the interfacial mobility and the local fluid velocity compared to clean interfaces, measurement with SFV and calculated Marangoni Stress by the velocity gradients from inside and outside of droplet interfaces by Hosokawa et al. [Hos14b] [Hos17]

Overall surfactants effects on transport processes at bubbles can be measured easily but local quantitative data to verify the existing models have been a big challenge for experimental researchers until Hosokawa et al. made surprisingly simple experiments with oil droplets in contaminated water, measuring the local velocities of both fluid phases with Spatial Filter Velocimetry (SFV). They have shown the different regimes of circulation flows, stagnant cap angles depending on the surfactants concentration (Fig. 2.5a ) and they have calculated the Marangoni stress from the local velocity gradients at both sides of the interface (Fig. 2.5b) [Hos14b] [Hos17]. Unfortunately up to now, no comparable experiment proving the immobilization of interfaces with such a high accuracy for bubbles exists. The additional acquisition of the local gas velocity data inside a bubble is not accessible via typical experimental approaches.

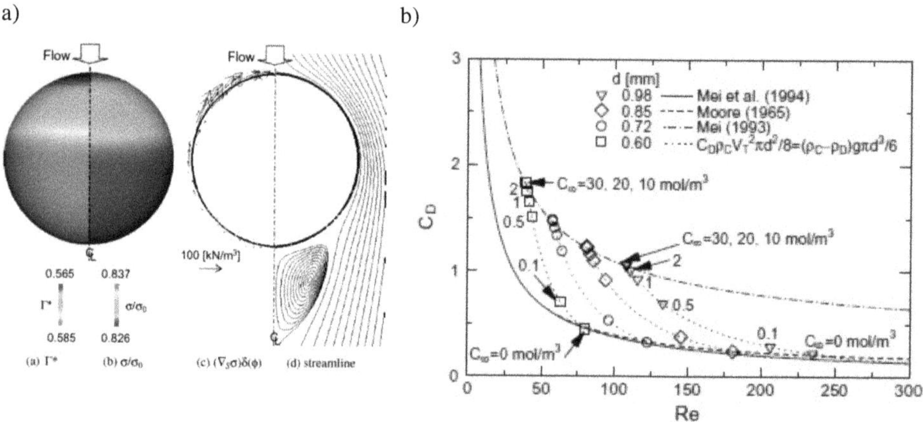


Fig. 2.6 Effects of surfactants on interfacial mobility and drag coefficient of rising bubbles - a) Interfacial TritonX-100 density $\Gamma$ and tension ratio $\frac{\sigma}{\sigma_0}$ of a rising spherical bubble ($d_b = 0.72$mm; $C_{Triton,\infty} = 30$mol m$^{-3}$) and its resulting Marangoni force; b) Effects of surfactants onto the drag coefficient of rising air bubbles [Hay12]

Hayashi et al. have shown before, that surfactants can reduce the rise velocity of bubbles due to their influence on the boundary condition at the interface which is influencing the local fluid velocity. Therefore, the deposition of streamlines from the interface can be shifted, such that wake structures can change from creeping laminar flow to toroidal vortex as seen Fig. 2.6 a)), which is depending on the interfacial surfactants density $\Gamma$. Both effects together are influencing the drag coefficient of bubbles as seen in Fig. 2.6 b) (compare with Fig. 2.2 and Fig. 2.3). With increasing surfactant concentration in the bulk $C_\infty$, more surfactant molecules are adsorbed and shift the equilibrium of adsorption/desorption at the interface, therefore the drag is developing from a clean interface to rigid interface value. In Fig. 2.6 b) bubble Reynolds numbers are reduced, due to increasing $C_D$, caused by increasing surfactants concentration $C_\infty$. For contaminated bubbles $0.6 \leq d_B \leq 2$ mm this effect can have significant influence, which reduces their rise velocity by a factor of 2 compared with bubbles in pure water, compare Fig. 2.3. In summary, the effect of surfactants on the fluid dynamics can not be generalized and depends strongly on the fluid system and the surfactants properties and

concentration, as well as on the local acting shear flow, which induces the counter acting Marangoni force. But in the presence of surfactants the rise velocity of free rising bubbles is reduced and therefore this value can be used as reference for the gas-liquid systems, if the impurity concentration is so low, that no mayor influence can be measured. This does not necessarily mean that the multiphase system is pure, but at least it is an indicator.
In general, there are different approaches for universal flow maps of free rising bubbles to summarize all the occurring bubble regimes, depending on the fluid properties. Clift et. al and Tripathi et al. used a method combining representative dimensionless numbers to characterize the occurring bubble shape and rise regimes as seen in Fig. 2.7 a) & b) [Cli78] [Tri15]. Both flow regime maps show, that actual spherical bubbles are just occurring at a small area of property ranges in this logarithmic maps. All the other occurring bubble shapes show irregular shape deformation and highly stochastic rise behavior. Therefore, small spherical bubbles are a good start for experiments but it is challenging to transfer the findings into other application relevant regimes and even to get reliable reproducible experimental data for those bubbles.

Before continuing with wall effects on fluid dynamics of bubbles, the mayor established mass transfer models at simplified spherical bubbles will be introduced in the following chapter.

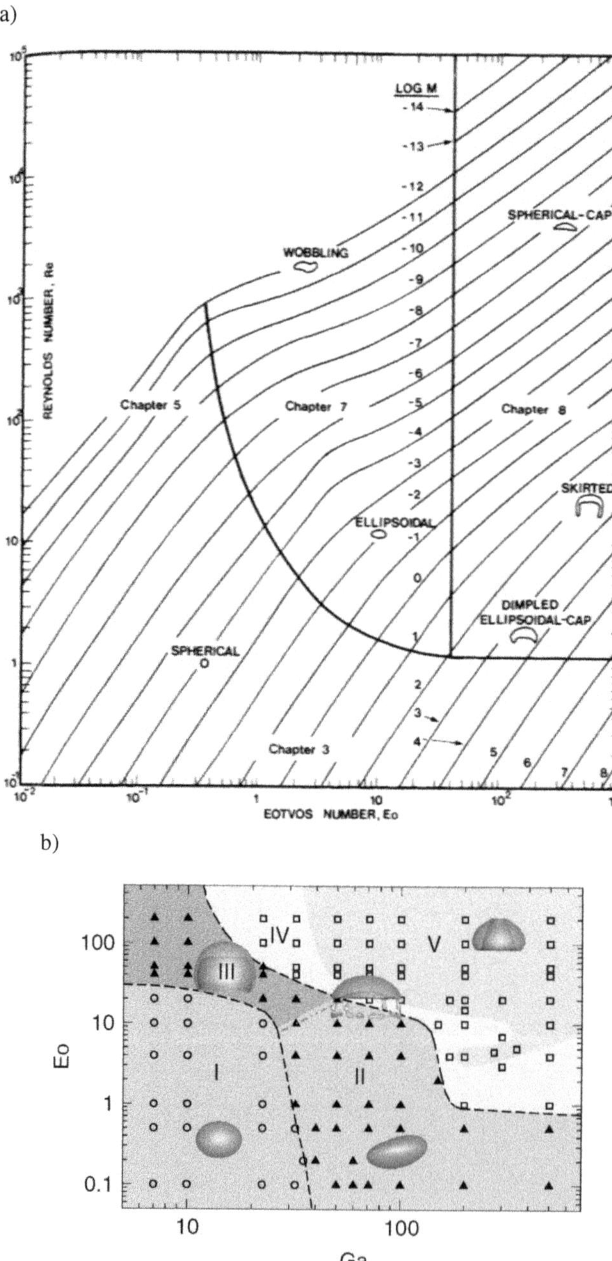


Fig. 2.7 Shape and velocity regimes of rising bubbles. a) Bubble shape regimes depending on their rise velocity and fluid properties (*Re* vs. *Eo*; lines of constant *Mo*) [Cli78]; b) Bubble shape regimes defined only by fluid properties (*Eo* vs. *Ga*) [Tri15]

### 2.2.2 Mass transfer at free rising bubbles

Mass transfer of species from a dispersed phase into a continuous bulk phase is a complex interlinked process within both phases depending on interfacial effects. Therefore, simplified models are generally used to describe the processes, which are briefly introduced in the following. Additionally, the quantitative application of the models on $CO_2$ transfer from a dispersed bubble into aqueous solution is explained.
The transfer of a species inside one fluid is usually seen as a passive scalar, where the species is inert regarding fluid dynamics, as long as the fluid properties remain constant [Pop00]. The specie is transported with the flow and is balanced like

$$\frac{\partial C_i}{\partial t} + \nabla \cdot (C_i \vec{u}_j) = (\nabla^2 \cdot C_i) D_L \tag{2.13}$$

$C_i$ is the concentration of the species $i$, $t$ depicts the time, and velocity vector $\vec{u}_j$ with its three dimensions $j = x, y, z$ (or the components $u, v, w$) in a diluted system, where a reaction is neglected in this context for simplification. In Eq. 2.13, the first summand on the left side depicts the unsteady term, the second one the convective term, and on the right side the diffusive term. The diffusive term consists of the derivative of a concentration gradient multiplied by the Diffusion coefficient $D_L$. The diffusion coefficient $D_L$ depends on temperature $T$ and concentration $C$, but can be considered to be constant when the concentration is low. The concentration gradient $\frac{dC}{dx}$ is the driving force of a diffusive flux $j$ within a single phase defined by the Fick's first law

$$j = -D_L \frac{dC}{dx}. \tag{2.14}$$

As soon as an interface separates two phases with different fluid properties regarding a species transfer it gets more complicated and simplified approaches have been established to model the observed and theoretical processes. The two main models will be introduced in the following.

#### Two Film Theory

For mass transfer across an interfaces simplified theories like the Two-Film Theory have been established [Lew24]. This theory assumes that in unbalanced gas-liquid systems thin boundary layers are formed on both sides of the interface, as shown in Fig. 2.8. In both films convective transport is negligible and only molecular diffusion according to Fick's law (Eq. 2.14) is taking place. Here the velocity and concentration gradients of a compound $i$ act only along the normal direction of the interface (here $x$). Flow conditions are assumed to be stationary as well as laminar and any occurring mass transfer is due to diffusion. The interface is defined without any resistance, capacity or dimension in normal direction. This defines all transport resistances, for the transfer of species $i$ from one phase into another, to occur inside the films connected in series. Phase equilibrium can be considered at the interface and Henry's law is applicable and defines the solubility constant

$$H^* = \frac{C^*_{d,i}}{C^*_{c,i}}. \tag{2.15}$$

where $C^*_{c,i}$ is the saturation concentration within the continuous phase, and $C^*_{d,i}$ within the dispersed phase [San15]. The concentration ratio of the species directly at the interface is called Henry jump. Figure 2.8 shows the Henry jump as well as the concentration progresses for the specific application. In case of an absorption of pure dispersed gas into a continuous liquid, the film thickness $\delta_d$ on the gaseous side can be neglected in comparison to the film thickness on the liquid side $\delta_c$. The ratio of fluid viscosity and therefore diffusion of the specie inside the phases is high, which leads to high difference in concentration gradients. In conclusion the mass transfer resistance is dominant on the liquid side. Additionally, it is assumed that the gas remains pure during the mass transfer process. The index expresses the phase, $c$ stands for continuous phase and $d$ for dispersed phase. The upper index $*$ means that a state at the interface is considered, and $\infty$ depicts values far away from the interface.

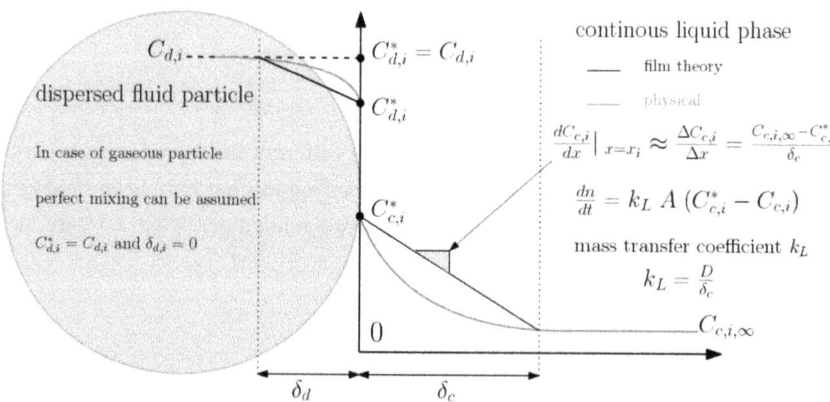


Fig. 2.8 Schematic representation of the Two Film Theory. Mass transfer of compound $i$ from a dispersed fluid particle into the continuous phase, according to Kraume [Kra17]. Concentration profiles according to the theory (black) and to the physical transfer (red) are plotted.

The Two Film Theory postulates that there is no convection within the film and that all species are transferred across the film by diffusion into the bulk phase. This enables to describe the amount of transferred species by the flux from Eq. 2.14. If the only species to be transferred is e.g. $CO_2$ ($i = 1$), the amount of transferred specie through the film $\delta_c$ (at $|_{x=x_1}$) is equivalent to the total transfer by mass continuity. With the simplification

$$\frac{dC_c}{dx}|_{x=x_i} \approx \frac{\Delta C_c}{\Delta x} = \frac{C_{c,\infty} - C^*_{C_c}}{\delta_c} \tag{2.16}$$

and the following applies

$$\frac{dn}{dt} = k_L \, A \, (C^*_c - C_{c,\infty}) \tag{2.17}$$

The mass transfer coefficient $k_L$ is a measure for the potential of the amount of transferred species per time across the total interfacial area $A$, where the flux is provided by the driving force $\Delta C$. Therefore, the mass transfer coefficient is defined as

$$k_L = \frac{D_L}{\Delta x} \approx \frac{D_L}{\delta_c}. \tag{2.18}$$

The dimensionless parameter which describes the mass transfer is the already mentioned Sherwood number

$$Sh_B = \frac{k_L \cdot d_B}{D_L} \tag{2.19}$$

where the characteristic length $L$ is e.g. the bubble diameter $d_B$. The Sherwood number is a function of the Reynolds and the Schmidt number,

$$Sh_B = f(Re, Sc), \tag{2.20}$$

with the Schmidt number being the ratio of the momentum and mass diffusivity. Applying the Two Film Theory, the Sherwood number remains as

$$Sh_{B,loc.} \approx \frac{d_B}{\delta_c} \tag{2.21}$$

from Eq. 2.18 and 2.19 as a simplified approach for local Sherwood number [1]. The experimental estimation of $Sh$ by this relation between $d_B$ and the local $\delta_c$ is challenging, because the boundary layer thickness depends strongly on the local velocity at the bubble, which depends on the bubble rise regime, that usually shows time dependent dynamic changes. But if this value is accessible by local optical measurement techniques, than $Sh_{loc}$ can be calculated, which enables a deeper insight into the interlinked transfer processes.

[1] After discussion with Prof. Akio Tomiyama

### Penetration or Surface Renewal Theory

Another important mass transfer theory is the Penetration Theory [Hig35], which was further modified by Davies to the Surface Renewal Theory [Dav50]. Both do not involve the concept of films but in-stationary processes in fluid elements during their surface contact times. In Fig. 2.9 fluid elements of the turbulent continuous phase, with initial species concentration $C_{c,i}(t = t_0)$ of $i$, are reaching the interface, remain for a certain residence time $\tau$, meanwhile the concentration $C_{c,i}(t > t_0)$ of $i$ increases, until the fluid element is leaving the interface and transporting the absorbed species $C_{c,i}(t_1)$ into the bulk of the continuous phase.

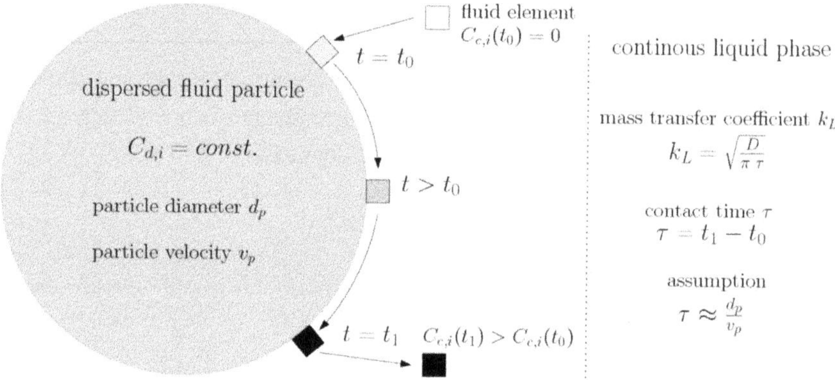


Fig. 2.9 Schematic representation of the Penetration Theory. Mass transfer of compound $i$ from a dispersed fluid particle into an fluid element of the continuous phase. Concentration $C_{c,i}$ of fluid elements is increasing depending on the contact time $\tau$.

Therefore, the residence time distribution of the interacting fluid elements is the key issue of the application of the theory and to estimate the mass transfer coefficient of the fluid system. A common assumption is to describe the contact time $\tau$ by depicting the renewal rate of the liquid surface by the ratio of the bubble diameter $d_B$ and its rise velocity $v_B$. Then, the mass transfer coefficient is

$$k_{L,SR} = \sqrt{\frac{D_L}{\pi\,\tau}}. \tag{2.22}$$

The approaches of the simplified models of the Two Film and Surface Renewal Theory are contradicting. However, both can be valid approximations in contrary conditions as Toor and Marchello demonstrated [Too58]. Thus, the surface renewal theory should be applied for high Schmidt numbers and small residence times $\tau \ll \delta^2/D_L$, while the film model should be applied for low Schmidt numbers and higher residence times $\tau \gg \delta^2/D_L$. Anyhow, in the following determination of mass transfer the good applicability of both theories to establish the common empirical Sherwood correlations as a function of Reynolds and Schmidt number will be shown.

### Determination of mass transfer

Typically, the overall mass transfer performance is determined by experimental investigations, where the Two Film Theory is generally applied. The transferred mass $\dot{n}$ within a reactor can be defined as

$$\dot{n} = k_L{\cdot}a\left(C^*_{c,i} - C_{c\infty,i}\right) \tag{2.23}$$

according to Eq. 2.17. Based on this description the mass transfer coefficient can be experimentally determined for bubbly flows in a gas-liquid contact apparatus, if the specific interfacial area $a = A/V$, the total interfacial area $A_{tot}$ within a certain control volume or reactor volume $V_{reactor}$ is known. Usually $a$ is estimated for bubbly flows by the total interfacial area $A_{tot}$ of a theoretical bubble number with a measured mean bubble diameter $d_{32}$, called Sauter diameter, with a total summed bubble volume, which is equivalent to the total gas volume $V_g$ within the control volume $V_{reactor}$. Therefore, $a$ can be estimated by an approach as

$$a = \frac{6 \cdot \varepsilon_g}{d_{32}}, \tag{2.24}$$

where $\varepsilon_g$ is the reactors gas hold up. But for single bubble experiments $a$ can be measured directly, if the shape is steady and rotationally symmetric.
The non-dimensional relation of convective and diffusive mass transfer is the Sherwood number (Eq. 2.25), which is used for the description of the mass transfer performance in bubbly flows [Bra71].

$$Sh = \frac{k_L\, d_B}{D_L} \tag{2.25}$$

Most empirical correlations for the Sherwood number are based on the Reynolds (Eq. 2.2) and Schmidt number $Sc$

$$Sc = \frac{\nu}{D_L}. \tag{2.26}$$

The Schmidt number is the ratio of viscous diffusion rate to molecular diffusion rate and can therefore also be seen as relative ratio of the fluid boundary layer $\delta_{fl}$ and the concentration boundary layer $\delta_c$.
But the application of this definition to describe the transfer of species in a multiphase system is very difficult, due to complex 3D multiscale problems. During a bubble rise and

dissolution of species into a continuous fluid, neither the local information of $\delta_c$ nor even the exact shape and interfacial area of a single bubble are usually accessible. In realistic processes in multiphase reactors with high gas hold ups and turbulent flow structures it gets even more complex. Therefore, empirical Sherwood correlations are established using easy measurable global averaged process parameters to approach the physical processes as good as possible for certain Reynolds and Schmidt number regimes [Bra71], [Kra17], [Cli78].

The dependency of the Sherwood correlation can also be explained by using the introduced two mass transfer models with assumptions and simplification as follows

$$-D_L \frac{\partial C}{\partial x} \approx -D_L \frac{\Delta C}{\Delta x} = k_L \cdot \Delta C \tag{2.27}$$

with $\Delta x = \delta_c$ the expression

$$\frac{k_L \cdot \delta_c}{D_L} \tag{2.28}$$

becomes constant, because it defines the mass transfer coefficient from Eq. 2.18. The diffusion coefficient with its units [$m^2\ s^{-1}$] has to overcome the concentration boundary layer $\delta_c$, therefore $\delta_c$ can be used here as length

$$D_L \approx \frac{l^2}{t} = \frac{\delta_c^2}{t} \tag{2.29}$$

From the Surface Renewal Theory it is known, that the contact time $\tau \approx d_B/v_B$ and with substitution into Eq. 2.29 it follows

$$D_L \approx \frac{v_B \cdot \delta_c^2}{d_B} \tag{2.30}$$

$$\delta_c \approx \sqrt{\frac{D_L \cdot d_B}{v_B}} \tag{2.31}$$

Taking the Sherwood definition and extending with a length ratio $d_B/\delta_c$ and substitution of Eq. 2.30 and Eq.2.28

$$Sh = \frac{k_L \cdot d_B}{D_L} = \frac{k_L \cdot \delta_c}{D_L} \cdot \frac{L}{\delta_c} = C_1 \cdot \frac{d_B}{\delta_c} \approx C_1 \cdot \sqrt{\frac{v_B \cdot L^2}{D_L \cdot d_B}} \tag{2.32}$$

the expression in the square root equals the definition of Peclet number $Pe = Re \cdot Sc$

$$Sh \propto C_1 \cdot Pe^{1/2} = C_1 \cdot Re^{1/2} \cdot Sc^{1/2\ or\ 1/3} \tag{2.33}$$

Most empirical Sherwood correlations have this similar structure and are depending on Reynolds and Schmidt number [Bra71][Cli78][Kra17] and are typically of the form

$$Sh = 2 + C_1 \cdot Re^n \cdot Sc^m \tag{2.34}$$

where the summand 2 is the theoretical solution for a spherical gas bubble at rest in a stagnant liquid and depicts the limit of transfer if only diffusion takes place [Kra17]. $C_1$ is a factor that is depending on the Schmidt number and on the fluid viscosity ratio. The exponents $n$ and $m$ and the factor $C_1$ are empirically determined and used as fitting parameters to adapt the correlation to the slope of experimental values and where other unknown or non measurable influences like shape deformation are taken into account.
This comparison shows that both main mass transfer theories have their eligibility for spherical bubbles and general empirical Sherwood correlations are using the physical dependencies with a few assumptions and corrections. But it is noted, that usually these empirical correlations are only valid in a specific range of operating conditions and of the exact gas liquid system and the geometry of the contact apparatus. For sure, this is caused by the lack of knowledge of the local conditions at every single bubble or fluid element of the bubbly flow. But local investigations at simplified bubbly flows like single bubbles help to understand the effects of local processes to the overall process, which helps to improve CFD methods for reliable process design.

**Influence of surfactants on mass transfer**

As mentioned before in Chapter 2.2.1, surfactants influence the fluid properties and also the local fluid dynamics of rising bubbles, by accumulation at interfaces, where they form molecular layers. Therefore, it is not surprising that those molecular layers at interfaces are affecting the mass transfer performance across this interface by an additional diffusive resistance. Since, nearly every system includes contaminations or impurities, especially aqueous systems, it is questioned "which dose makes the poison" and which type and concentration of surfactants has a significant influence on transport processes that can be measured and identified in a gas-liquid system.

Due to the shear forces during the bubble rise, adsorbed surfactants are transported downwards and accumulate at the bubble rear, where they can be desorbed into the continuous

fluid again. The Maragoni force works against the shear and tends to distribute surfactants uniformly across the interface, to compensate the imbalance of surface tension [Gri60] [LM65] [Duk16] [Hos17]. This can reduce the internal circulation of the bubble and the local fluid velocity of the continuous fluid at the interface. Lower velocities lead to lower convective transport, which affects the boundary layer thickness and residence times for the mass transfer theories, as introduced before in Chapter 2.2.2. Additionally, rigid caps by accumulated surfactants change the local fluid flow by earlier deposition of streamlines from the bubble interface, and may create wake structures with re-circulation flow Fig. 2.6 a). This reduces the area of the bubble which is in direct contact to the liquid bulk phase ($C_\infty$), which would ensure a higher concentration difference $\Delta C$, while specie accumulation in the wake reduces the same locally. Therefore, the reduction of local mass flux leads to a reduction of integral mass transfer across the bubble interface into the continous phase. Those effects of interfacial contamination on free rising bubbles have been investigated recently by means of numerical simulations [Jia17] [Pes18]. With increasing surfactants concentration, the mass transfer is lowered, which has been shown also by means of experiments carried out by Rosso et al. [Ros06]. They propose a dimensionless number for the characterization of surface contamination, which is then used for a Sherwood correlation taking into account the influence of surfactants on mass transfer empirically. According to Pesci et al., reaching an equilibrium of surfactants adsorption and desorption is not required for reaching a quasi stationary bubble rising behavior. Therefore, a terminal bubble rise velocity is not an indicator for interfacial surfactants equilibrium, which makes investigations on reason and impact of surfactants affecting transport processes even more difficult. This overview of transport processes at free rising bubbles gives the basic knowledge of the interconnection between transport processes in gas-liquid flows, which is transferred to bubbles in channels in the following.
In summary, the mass transfer is influenced in general by:

- Bubble shape deformations leads to time depending interfacial area, therefore the boundary layers $\delta_{fl}$ & $\delta_c$ are stretched in tangential direction and their thickness in normal direction is decreased, which usually enhances the transport in normal direction.
- Transport of other species in the system, which affects the mole fractions in the bubbles, which reduces the transport driving force.
- Impurities or additives like surfactants, accumulated at the interface and can act as an additional mass transfer resistance and lower the interfacial mobility, therefore increases the boundary layer thickness $\delta_c$, which lowers the transport.

### 2.2.3 Wall effects on rising bubbles in vertical channels

In the preceding section analytical solutions and numerical and experimental data of the fluid flow around moving bubbles have been introduced, where only the flow in the vicinity of the interface is affected, while the bulk phase flow stays uniform. As soon as solid walls are in the vicinity of rising bubbles, the boundary conditions of the continuous phase are changing to no-slip and fluid boundary layers with velocity gradients are generated. Local velocity gradients cause pressure differences and therefore can effect the interfacial shape and the rise behavior of bubbles. Clift et al. have shown, that in dependency of the distance between walls and particles or bubbles, these effects on the flow profile of the continuous phase (Fig. 2.10 a)) can also have dominant interactions with the bubble, e.g. reducing their terminal rise velocity as shown in Fig. 2.10 b) [Cli78]. In this research the wall effect is induced by rotation or axial symmetric channel walls of pipes or ducts, with circular or square cross sectional areas. Effects on moving particles or bubbles in the vicinity of single walls are also very important for e.g. swarm behaviors and classification of bubbles near walls by asymmetrical lift forces induced by velocity and shear gradients in bubble columns. Fundamental research on lift force has been done by e.g. Tomiyama, Lucas et al. and Aoyama et al. [Tom02] [Luc11] [Aoy16] but is out of focus in this research work.

#### Flow regimes of small, intermediate and Taylor bubbles in channels

Clift et al. have shown in Fig. 2.10 a) the general coordinates and parameters of a model of a fluid particle moving in a vertical channel, where the diameter ratio $\lambda = d_p/D$ of the particle diameter $d_p$ and channel diameter $D$ is defined. In Fig. 2.10 b), the wall effect on the rise velocity

$$v_p/v_{p,free} = [1-\lambda^2]^{3/2} \tag{2.35}$$

is shown as velocity ratio of a wall effected fluid particle $v_p$ to a free rising one $v_{p,\infty}$. The rise velocity is unaffected for small diameter ratio $\lambda < 0.1$, but strongly decreasing with increasing $\lambda$. The model shows a good agreement for fluid particles, with a diameter ratio $\lambda < 0.6$. However, the data for larger particles can show strong deviation from this equation, by e.g. dynamic shape oscillation and deformation of bubbles or droplets in comparison of free rising ellipsoid bubbles. The range of applicability is given by $Eo < 40$, $Re < 200$ and $\lambda < 0.6$. For a smaller $Eo$ numbers the influence of the interfacial tension becomes stronger, which enables more stable bubble shapes and reduces interfacial waves.

With increasing diameter ratio and therefore wall effect, those droplets and bubbles tend to deform from spherical to ellipsoids to bullet shaped gas slugs and the particle diameter $d_p$ needs to be calculated by a sphere volume equivalent bubble diameter to stay comparable.

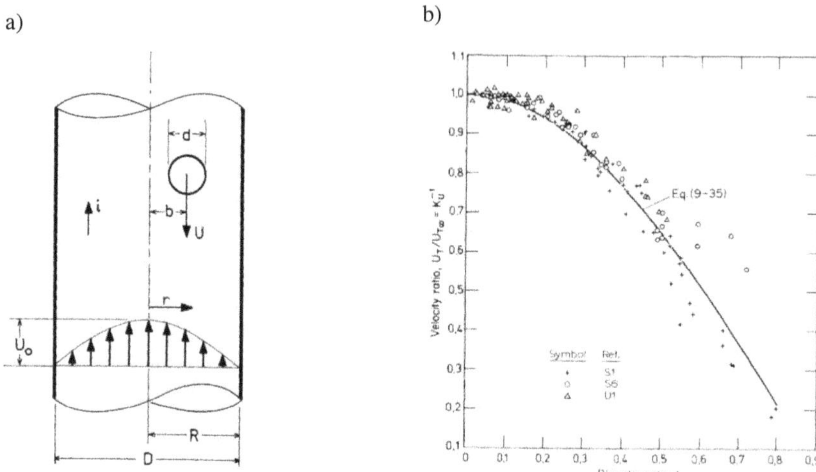


Fig. 2.10 The effect of the vicinity of solid walls on rising bubbles. a) Schematic coordinates of a fluid particle in channel. b) The influence of the diameter ratio $\lambda = d_p/D$ of particle and channel onto the particle velocity in a vertical channels [Cli78]

The rise velocity of fluid particles $0.5 < \lambda < 0.8$ differ even stronger from the model but unbounded droplets and bubbles in this range tend to be ellipsoid and flattened in vertical direction, while wall effects tend to elongate the bubbles. These elongated bubbles develop to gas slugs, when $\lambda \gtrsim 0.8$, which fill nearly the hole channel cross section despite a thin liquid film between bubble and channel wall as seen in Fig. 2.11.

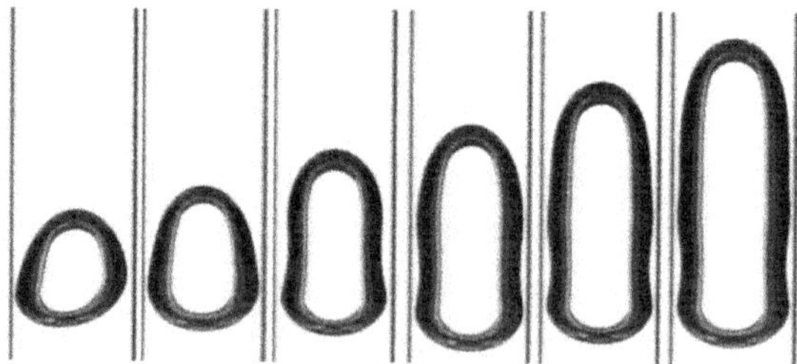

Fig. 2.11 Example bubble shapes caused by wall effects for $\lambda > 0.8$ rising bubbles in a vertical channel $D = 7$ mm ($Eo$ = 6.7)

The bubble shape depends strongly on the $Eo$ and the bubbles shown represent only gas bubbles in water filled channel $D = 7$ mm and different shapes can be expected at other fluid systems. In any case, the deformation to elongated gas slugs influences the rise velocity as e.g. Maeda showed for rising bubbles (Fig. 2.12a) or Kurimoto et al. for droplets (Fig. 2.12b)

[Mae75] [Kur13]. In Fig. 2.12a) it is also shown, that bubbles in channels with a diameter below some critical value will not rise anymore, if their $d_{eq}$ tend to reach $D$, as shown for data in $D = 2.3$ and 4.9 mm. In larger channels the bubble and droplet rise velocity for $\lambda \approx 1$ start to flatten until they reach constant values for $\lambda \gg 1$ independent of their particle size or volume. In general, this is the characterization of flow regimes in channels, with small, intermediate and large bubbles, which have a volume independent rise velocity. The latter are called Taylor bubbles after G.I. Taylor. This significant behavior of Taylor bubbles is totally different to the rise behavior of free rising bubbles and unique for two phase flows.

When the sphere volume equivalent fluid particle diameter becomes larger than the channel diameter $D$, the fluid particles elongates and gets a bullet shape, which is called slug flow or Taylor bubble or drop. The advantages and resulting possibilities of Taylor bubbles as a special research field in multiphase flows will be discussed in the next chapter.

## 2.3 Taylor bubbles

Investigations on the rise behavior of elongated or bullet shaped bubbles in vertical and liquid filled channels have started with Gibson in 1913 [Gib13] and raised interest between the 50s and 70s, when fundamental studies on this system have been published and universal correlations on rise velocities have been established by Dumitrescu [Dum43], Davies and Taylor [Dav50], Uno [Uno56], Taylor [Tay61], Bretherthon [Bre61], White and Beardmore [Whi62], Brown [Bro65], Wallis [Wal69], Maeda [Mae75] and Tung [Tun76]. When exactly the honor of naming elongated bubbles as "Taylor bubbles" after G.I. Taylor, who actually was concentrating on problems using simple equipment after his retirement in 1952, has been started is not known. The Taylor bubble has been identified to be an extraordinary two phase flow tool for the investigations of fluid dynamics and mass transfer processes and the effects of fluid properties. The Renaissance of fundamental research on Taylor bubbles started in the end of the $20^{th}$ century, when methods for numerical simulations rose up and Taylor bubbles became an promising two phase flow system for systematic validation of numerical methods focussing on fluid dynamics, mass transfer processes and effects of impurities [Tom96] [Via03] [Hay11] [Kur13] [Mor16]. The evidently reduced complexity of Taylor bubbles led to a large data base on experimental results and improved numerical methods to take into account fluid properties and thereby the influence surfactants and electrolytes [Aok13] [Aok15] [Hor16] [Hor19] [Hor20]. Among others, by the big effort of the research group of Prof. Tomiyama accelerated this research field.

a)

b)

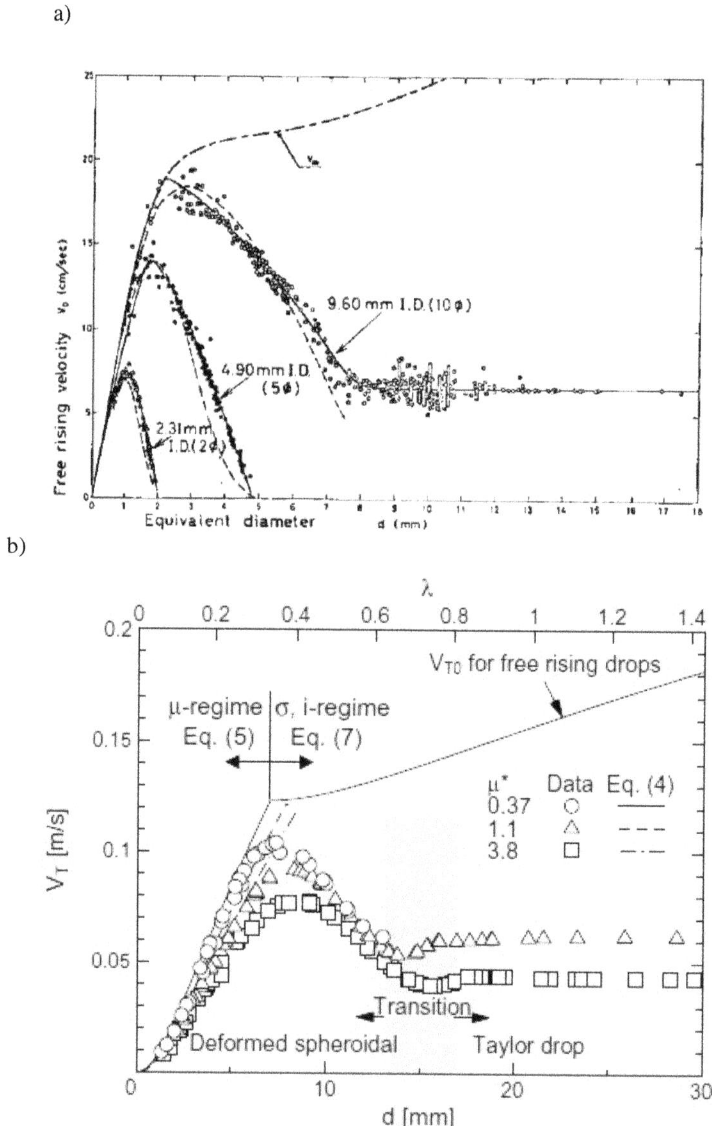


Fig. 2.12 Influence of channel walls onto rising fluid particles in various channels $v_b$ vs. $d_{eq}$. a) Velocity ratio of bubbles rising in vertical channels $D = 2.3, 4.9$ and 9.6 mm compared to free rising ones (dashed line). Influence of the equivalent bubble diameter on the rise velocity [Mae75]; b) Drop rise velocity in vertical channel. Influence of the viscosity and the droplet size onto the rise velocity [Kur13]

Actually, Taylor bubbles can also be found in various applications with a large range of channel sizes, like microfluidic channels in lab-on-a-chip devices [vB04] [Kre05] for pharmaceutical or special chemistry process engineering, in millimetric scale [Ala13] [Mey14] [But16] [But18] for e.g. gas-liquid-flows in monolith reactors in chemical engineering and catalysis chemistry. As introduced above Taylor bubbles appear in centimeter scale like pipes and channels from heat exchangers or pipe reactors. Already in 1913, Gibson published, that depending on the pipe or channel size, the different forces can change their dominance in the system, which effects the bubble shape, the interface stability and bubble rise velocity. Therefore, researchers from different fields are concentrating on Taylor bubbles and study the global and local processes in various fluid systems. Many researchers found a critical channel diameter, where buoyancy driven Taylor bubbles are separated from not rising bubbles, which stick inside the channel without a pressure driven flow. This is caused by the buoyancy force counteracting interfacial tension of the fluids, where the non-dimensional ratio of both forces is expressed with the Eötvös number $Eo$. In small channels the interfacial tension is strong enough to press the interface to the channel wall and reduce the liquid film thickness, where friction becomes dominant and the rise of bubbles is restrained. This work focus on buoyancy dominant Taylor bubbles rising spontaneously in liquid filled vertical channels ($Eo > 4$).

### 2.3.1 Terminal rise velocity of Taylor bubbles

Since Gibson described the volume independent rise velocity of Taylor bubbles in 1913, many researchers investigated the effects of channel diameter, channel cross section and fluid systems with the goal to establish universal empirical correlations with a wide range of applicability. White and Beardmore [Whi62] established a flow map with flow regimes of Taylor bubbles in various liquids in 5 mm $< D <$ 38.7 mm channels and in comparison with literature values, as seen in Fig. 2.13 a). Taylor bubbles a required to have a higher Eötvös number than the critical value $Eo > 4$ before they will rise by buoyancy. With increasing channel diameter $D$ the rise velocity or here in Fig. 2.13 the Froude number $Fr$ increases. Additionally, they indicated certain areas, where Taylor bubble velocities are independent of certain fluid parameters Fig. 2.13 b). This already indicates how complex the coupled processes at Taylor bubbles can be and why universal empirical correlations are difficult to be established, even if they look very simple at first glance. Clift et al. combined the maps shown in their textbook. Meanwhile many researchers came up with similar correlations, as seen in Table 2.1, until in the beginning of this century. Hayashi et. al [Hay11] and Kurimoto et al. [Kur13] established correlations for Taylor bubble rise velocity by with the combination of experimental and numerical data, for an even better prediction in a wide range of dimensionless groups.

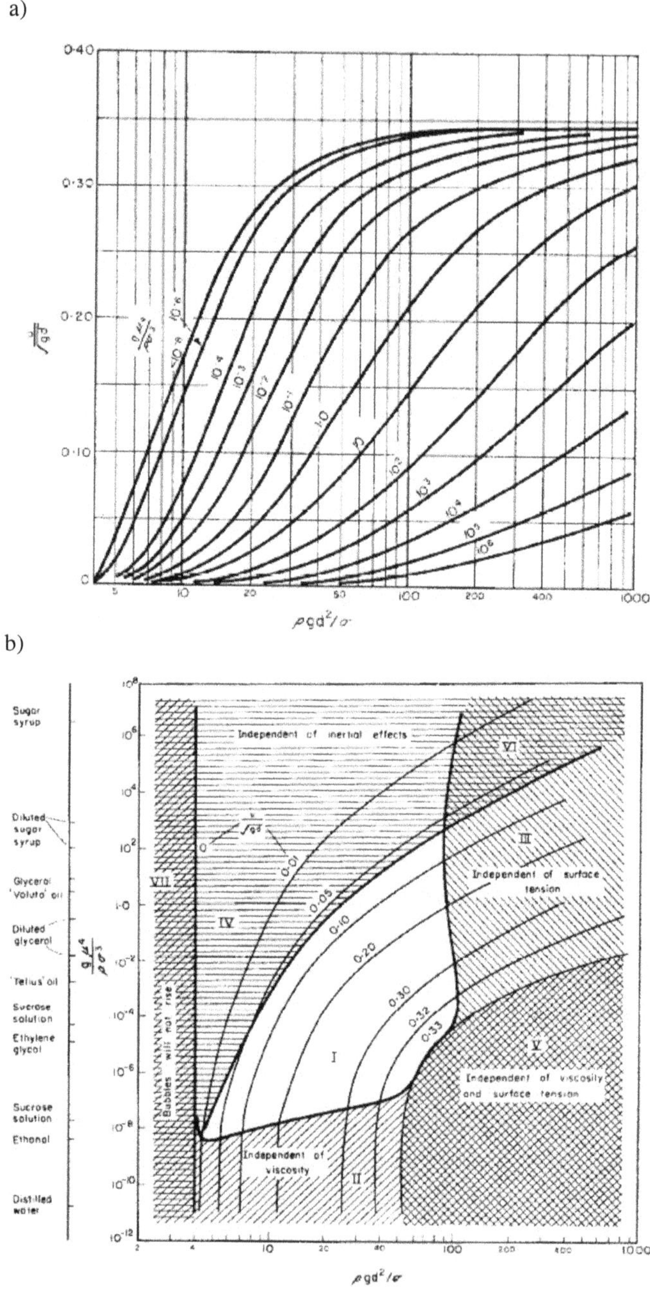


Fig. 2.13 Flow map of Taylor bubble rise regimes. a) Froude number vs. Eötvös number of Taylor bubbles established by various experiments and literature data; b) Independent regimes of Taylor bubbles rise velocity of certain fluid parameters. Property group over *Eo* with the indication of usual fluids. Note: *d* is the channel diameter in this figure. [Whi62]

Table 2.1 Summary of various correlations for the rising velocity of Taylor bubbles in a vertical channel according to Morado et al. [Mor16].

| Author(s) | Type of study | Correlation | Assumptions and comments |
|---|---|---|---|
| Dumitrescu (1943) | Theoretical [Dum43] | $v_B = 0.35\sqrt{g\,D}$ | • inviscid liquid<br>• constant upstream vorticity<br>• spherical nose<br>• negligible bubble expansion<br>• viscous and surfaces tension effects<br>• potential flow |
| Davies & Taylor (1950) | Theoretical [Dav50] | $v_B = 0.328\sqrt{g\,D}$ | • pressure over the bubble's nose equal to that over an ideal sphere<br>• negligible bubble expansion, viscous and surface tension effects<br>• potential flow |
| White & Beardmore (1962) | Experimental [Whi62] | $v_B = 0.345\sqrt{g\,D}$ for inertia-controlled regime graphical correlations for regimes controlled by different retarding forces | |
| Brown (1965) | Experimental & Theoretical [Bro65] | $v_B = 0.35\sqrt{g\,D}\sqrt{1-2(\frac{\sqrt{1-ND}-1}{ND})}$<br>$N = \sqrt[3]{14.5\rho_l^2 g/\mu_l^2}$ | • empirical application limits<br>Surface tension:<br>$\frac{\rho_l\,g\,D^2}{4\sigma}(1-2(\frac{\sqrt{1-ND}-1}{ND}))^2 > 5.0$<br>Viscosity: $ND < 60$ |
| Wallis (1969) | Theoretical [Wal69] | $Fr = \frac{v_B}{\sqrt{g\,D}}$<br>$-0.345(1-e^{-0.01\,N_f/0.345})(1-e^{(3.37-Eo)/m})$<br>$m = \begin{cases} 25, & N_f < 18 \\ 69\,N_f^{-0.35}, & 18 < N_f < 250 \\ 10, & N_f > 250 \end{cases}$ | |
| Tung & Parlange (1976) | Experimental & Theoretical [Tun76] | $Fr = \frac{v_B}{\sqrt{g\,D}} = \sqrt{0.136-0.944\frac{\sigma}{\rho_l\,g\,D^2}}$ | • Negligible viscous effects<br>• Isolated interfacial effects<br>• Excellent agreement with experimental data |
| Viana et al.(2003) | Experimental [Via03] | $Fr = \frac{v_B}{\sqrt{g\,D}} =$<br>$\frac{0.34/(1+3805/Eo^{3.06})^{0.58}}{(1+(\frac{R}{31.08}(1+\frac{778.76}{Eo^{1.96}})^{0.46})^{-1.45(1+\frac{7.22\times10^{13}}{Eo^{9.93}})^{0.094}})^{0.071(1+\frac{7.22\times10^{13}}{Eo^{9.93}})^{-0.094}}}$<br>$R = \frac{\sqrt{D^3 g\rho_l(\rho_l-\rho_g)}}{\mu_l} \sim N_f$ | |
| Hayashi et al.(2011) | Experimental & Numerical [Hay11] | $Fr = \frac{v_B}{\sqrt{gD}}$<br>$= \sqrt{\frac{0.0089}{0.0725+\frac{1}{Re_B}(1-0.11Re_B^0.33)}(1+\frac{41}{Eo^{1.96}})^{-4.63}}$ | • $10^{-7} < Re_B < 10^4$<br>• $4 < Eo < 3\times10^3$<br>• $10^{-2} < N_f < 10^5$<br>• $10^{-11} < M < 10^{10}$ |
| Kurimoto et al.(2013) | Experimental & Numerical [Kur13] | $Fr = \frac{v_B}{\sqrt{g\,D}} = \sqrt{\frac{G\,H}{\frac{H}{0.35^2+F}}}$<br>$G = (1+\frac{3.87}{Eo1.68})^{-18.4}$<br>$H = 0.0025[3+G]$<br>$F = \frac{1}{Re_B}(1-0.05\sqrt{Re_B})$ | |

### 2.3.2 Local fluid dynamics at Taylor bubbles: front, film & wake flow

Local fluid dynamics are more complex to be investigated experimentally than expected due to the small dimension of the fluid film between bubble and wall. For examination of the interface very high spatial resolution and a good optical entrance into the system with less refraction is required. But there are several researchers, who provided the first steps into this field. Taylor himself proposed analytical solutions of the velocity field around Taylor bubbles in channels [Tay61]. In 1988 Campos & Guedes de Carlvalho visualized the open and closed wake structures behind rising Taylor bubbles depending on *Re*. They managed to measure wake lengths for Taylor bubbles, which were rising through dyed fluid entering clear fluid, while the dye accumulates shortly in wakes, which enabled to characterize open and closed wake structures [Cam88] [Pin98]. Numerical simulations from Bugg et al. confirmed Taylor bubble shapes wake structures depending on $Eo$ of the system [Bug98] [Bug02]. Meanwhile, Particle Image Velocimetry (PIV) established accurate measurements of local fluid motion in multiphase systems by recording the displacement of tracer particles in the fluid and several researchers confirmed counter-rotating vortices or toroidal vortices behind rising Taylor bubbles [Bug02] [Van02] [Nog06]. Local investigations of the fluid dynamics in Taylor bubble train flows in small or micro channels have been carried out by Thulasidas et al. [Thu97] [Tso07]. In square channels King et al. [Kin07] and Meyer et al. [Mey14] found together with Aland et al. unexpected back flow regions in fluid films at contaminated interfaces [Ala13]. Recently, Butler et. al characterized the wake structures between Taylor bubbles of pressure driven Taylor flows, depending on the bubble distance for mixing reasons, but the flow conditions in micro channels are very different to those from Taylor bubbles in vertical channels rising by their buoyancy. Detailed velocity field measurements and flow characterization have been missing so far, for rising Taylor bubbles in vertical channels $5.5 < D < 10$ mm and $Eo < 10$.

### 2.3.3 Mass transfer at Taylor bubbles

#### Global Mass Transfer

Mass transfer at Taylor bubbles rising in various channel sizes have been investigated within the last two decades extensively by several research groups, where Prof. Tomiyama's research had big impact and influence on this research field [vB04] [Van05] [Abe08] [Abo13] [Aok13] [Hos14a] [Hay14] [Hos15]. Hosoda et al. investigated the dissolution of small and Taylor $CO_2$ bubbles in pipes and determined the mass transfer coefficient based on the rate of bubble shrinkage. With precise shadowgraphy from two directions they managed to reconstruct the bubble volume even for bubbles with slight shape dynamics and established correlations for the overall mass transfer which are valid for pipes with inner diameter of 12.5 mm $< D <$ 25 mm, as shown in Fig. 2.14.

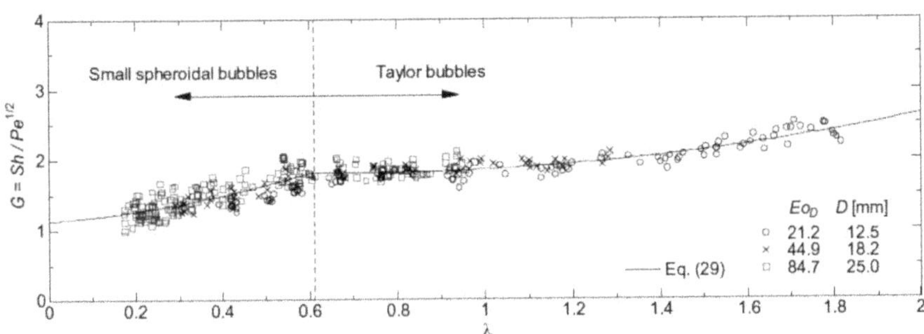


Fig. 2.14 Sherwood correlation for small, intermediate and Taylor $CO_2$ bubbles in vertical channels 12.5 mm $< D <$ 25 mm (20 $< Eo <$ 85) [Hos14a].

During the mass transfer process it is important to take into account the counter transfer of dissolved gas components (e.g. $O_2$ and $N_2$) in the water, which will transfer into the bubble and influence its mole fraction. Therefore, it is necessary to take into account the counter transfer of $O_2$ and $N_2$, when Taylor bubbles are rising in a liquid, which is in equilibrium with air. This is impressively shown in Fig. 2.15 a) by the comparison of experimental data and theoretical modeling of the remaining gas volume in the bubble. In comparison the development of the gas mole fraction of the bubble is shown in Fig. 2.15 b) which is in good agreement with experimental gas analysis.

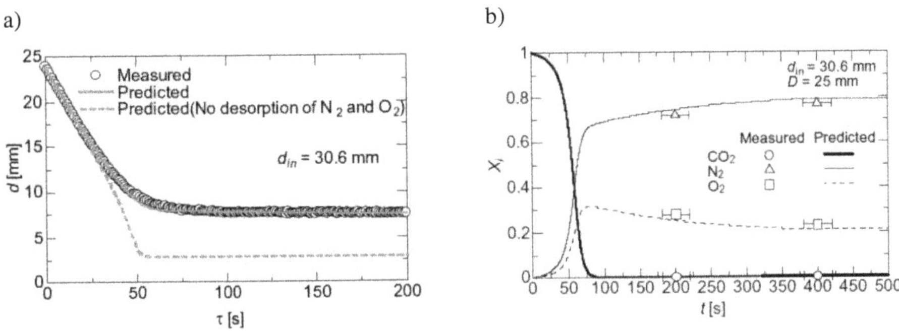


Fig. 2.15 Mole fraction change during longtime dissolution of $CO_2$ bubbles. a) The remaining volume of a Taylor bubble in equilibrium with water is changing due to counter current mass transfer from pre-dissolved gases like $O_2$ and $N_2$. $D = 12.5$ mm ; b) Comparison of simulated and measured gas mole fractions in bubbles during a long-term dissolution $D = 25$ mm. [Hos14a]

**Local mass transfer at the Taylor bubble interface**

The global mass transfer process, which was described in the preceding chapters, is the integral of all local mass transfer processes at the interface. The mass transfer is depending on the resistance, resulting from the local concentration gradient on the liquid side, where the mass flux is directly effected by the laminar layer thickness and from the actual $\Delta C$. The thickness of concentration boundary layer $\delta_c$ is affected mainly by the tangential velocity of the convective flow and the fluid dynamic boundary layer $\delta_{fl}$. The driving force $\Delta C/\Delta r$ is therefore additionally depending on the fluids element history and preliminary mass transfer into this element, compare Fig. 2.16 for local Sherwood number at the front $Sh_{FS}$ and rear $Sh_R$ of a Taylor bubble. Also there can be other mass transfer resistances directly at the interface, when contamination are adsorbed at the bubble, which will be discussed in the next chapter. Therefore, it is clear, that the local mass transfer and local Sherwood number $Sh$ can strongly differ in dependence of its position along the interface.

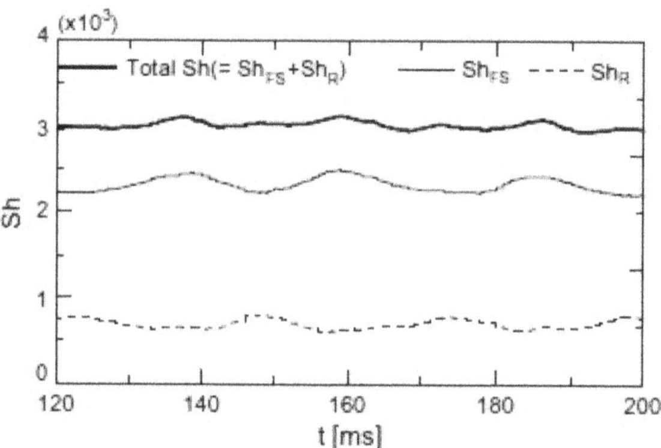


Fig. 2.16 Comparison between local Sherwood number of the front-side region $Sh_{FS}$ and the rear region $Sh_R$ of a Taylor bubble in $D = 12.5$ mm [Hay14].

### 2.3.4 Influence of surfactants on transport processes

Hayashi et al. showed via numerical investigations, how the equilibrium of interfacial surfactants distribution is depending on the surfactants properties and the local shear flow, compare Fig. 2.17 a), which causes locally different barrier effects by the surfactants molecules on mass transfer across the interface in normal direction. Additionally, they found an effect on the fluid streamlines, which lead to changed wake structures and therefore different mixing behind clean and contaminated Taylor bubbles. Intensive investigations of the influence of surfactants on global mass transfer rates, $k_L$, of single $CO_2$ bubbles in contaminated aqueous systems with alcohols in a vertical pipe of $D = 12.5$ mm diameter have been done by Hori et al., who identified the effect of the carbon chain length of the alcohols on $k_L$ [Hor16]. The four surfactants, 1-pentanol, 1- heptanol, 1-octanol and 1-decanol, were used to investigate their effects on ellipsoidal and Taylor bubbles. In dependency of surfactants concentration $C_{sol}$ they identified fully contaminated conditions of the bubbles and measured the corresponding rise velocities and mass transfer rates. In general the rise velocities and the mass transfer rates decreased with increasing surfactants concentration as expected from theory. Additionally, it was possible to identify the effect of the carbon chain length, while longer chain lengths cause stronger reductions, as seen in Fig. 2.17 b) and establish an extended $k_L$ correlation by taking into account the adsorption constants of the alcohols.

Additional to the effect of alcohols, Hori et al. recently investigated the combined effect of electrolytes e.g. NaCl and surfactants e.g. Triton X-100 [Hor17] [Hor19] [Hor20] on ellipsoidal and Taylor bubbles in the same experimental setup. They showed the reduction of the overall mass transfer rates due to the influence on the diffusion coefficient of $CO_2$ in aqueous system caused by NaCl. But NaCl shows an additional multiplier effect together with the surfactant Triton X-100 on mass transfer coefficient but not with the alcohols. The extraordinary data base, which is established in the research group of Professor Akio Tomiyama regarding fundamental experimental research on mass transfer processes enabled the validation of their in-house numerical methods for clean and contaminated multiphase systems including the surfactants distribution at bubble interfaces.

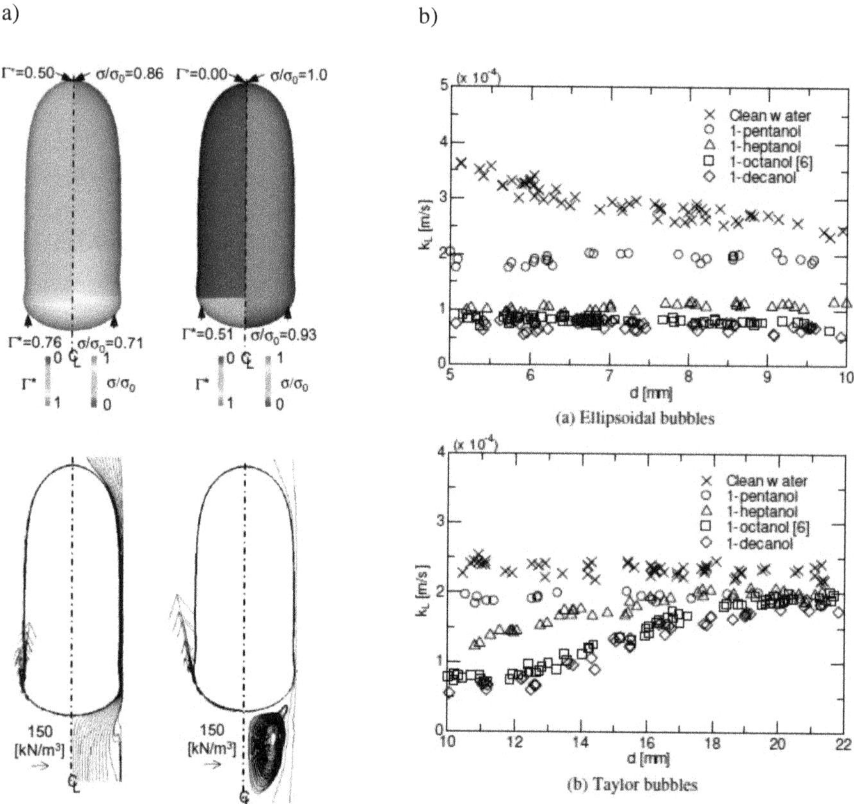


Fig. 2.17 Surfactants adsorption at a Taylor bubble in $D = 6$ mm channel and their effect on mass transfer coefficient $k_L$ in $D = 12.5$ mm channel. a) Adsorption of 1-pentanol and Triton X-100 their interfacial concentration distribution causing Marangoni force [Hay12]; b) Influence of various surfactants on the mass transfer coefficient $k_L$ [Hor16].

## 2.4 Conclusion, gap of knowledge & research goal

The fundamentals of bubbles rise velocity, drag coefficient and the most relevant mass transfer theories across gas-liquid interfaces have been introduced, as well as the behavior and effects of surfactants onto the transport processes. As soon as solid walls are in the vicinity of rising bubbles, their rise velocity is reduced due to shear effects between the bubble induced flow field and the wall. In vertical channels this effect increases with increasing diameter ratio $\lambda$ of bubble $d_B$ and channel diameter $D$ (wall distance) and can also influence the bubble shape. Therefore, bubbles at $\lambda \approx 1$ are becoming elongated while rising in vertical channels. These intermediate bubble regime shows a slightly volume dependent rise velocity. But the formation of a transition regime of this kind is strongly depending on the investigated fluid system and channel dimensions and not yet fully understood. The regime of elongated bullet shaped Taylor bubbles is defined by the beginning of the volume independent rise velocity, which is unique for distinguished gas-liquid system and enables fluid dynamically fixed bubbles due to rise velocity equivalent counter current liquid flow. Therefore, Taylor bubbles are very interesting for experimental investigations on the coupling of transport processes, because even long-term measurements with optical techniques can be established.

The classification of bubble flow regimes in vertical channels is based on the diameter ratio $\lambda$ of the sphere volume equivalent bubble diameter $d_{eq}$ and the channel diameter $D$, but the starting point of Taylor bubble regime depends on the fluid system and channel diameter. Until now, no universal correlation or physical key parameters have been identified to be able to predict the beginning of the Taylor regime. While the rise velocity of Taylor bubbles and therefore the *Re* number is depending on the channel diameter $D$, it has never been investigated in detail, how the local fluid dynamics at Taylor bubbles or the wake structures are depending on $D$. Hosoda et al. showed with a fluorescent dye the transition of the bubble wake structure from a dissolving Taylor bubble to small ellipsoidal bubble, which includes a drastically change in rise velocity or *Re* [Hos14a] [Hos15]. In comparison to rigid spheres, where the wake structure is a function of *Re*, characteristically wake structures with smooth creeping flow, stable closed toroidal vortex and open wakes with periodical vortex shedding have been found [Cli78]. Therefore, it follows to investigate Taylor bubble wakes and their local mixing intensity in dependency of their terminal rise velocity, *Re* number or channel diameter $D$.

Mass Transfer coefficient of Taylor bubbles is independent of the bubble volume and can be modeled by *Sh* correlations. Even the prediction of long-term dissolution is possible, if the counter diffusion of $N_2$ and $O_2$ is taken into account for Taylor bubbles in vertical channels $D > 12.5$ mm [Hos14a]. The applicability of these correlations to Taylor bubbles in smaller channels and therefore smaller $Eo$ numbers, where the influence of the interfacial tension becomes stronger, is an open research field.

Additionally, the possibility of visualization and measuring of local mass transfer processes at the interfaces, like concentration boundary layer, is higher due to the steady bubble shape for $Eo < 10$ with less dynamic shape oscillations and interfacial waves for the different Taylor bubbles and channel sizes. It would be interesting to compare experimental data with Prandtls boundary layer theory at plates and the analogy between transfer processes. With increasing velocity of the bulk phase, the fluid dynamic $\delta_{fl}$ and concentration $\delta_c$ boundary layers are becoming thinner, which should be accessible within Taylor bubble experiments.

The effect of surfactants on transport processes in multiphase system can be strong and depends on the interfacial concentration and the formation of molecular layers. Surfactants reduce for example the mass transfer coefficient and effects fluid dynamic boundary conditions by immobilization of interfaces, like it was shown by Hayashi et al. by numerical investigations for Taylor bubbles [Hay12] and by Hoskokawa et al. experimentally for liquid/liquid systems [Hos17]. This can lead to changes of fluid element paths called streamlines and therefore the wake structures behind moving fluid particles, which in general is depending on its $Re$ number.

To get better insights into the characteristics of bubble regimes in channels and the coupling of global and local transport processes with and without surfactants, new experiments under well defined conditions are needed. With a reduction of the Eötvös number steady bubble shape are more likely to be investigated, due to suppressed dynamical bubble shape oscillations. Coupling of global and local field of views and also of global and local measuring techniques are required to analyses and characterize the dependencies of fluid parameters and the Taylor bubble regime or the dependencies of local convection process and the concentration boundary layer $\delta_c$ at the interface. The applicability of existing mass transfer correlations to smaller $Eo$ would build the fundamental knowledge for detailed validations of local and global mass transfer investigations and a possible mass balance of the transferred species.

# Chapter 3

# Experimental set-up and procedure

The process performance of industrial multiphase reactors is usually controlled by adjusting key parameters like volume flow rates, system pressure or system temperature to set the optimal flow regime for a given reaction. Empirical correlations from experimental investigations help the engineers to establish stable regimes with e.g. high values for mass transfer coefficient $k_L$ or the specific interfacial area $a$ inside the reactor. Usual industrial reactors are not transparent and are made of stainless steel due to resistance and safety reasons. But optical entrance or transparency is still a key requirement for reliable and accurate measurements with a high spatial resolution. The direct measurements of the flow structures, bubble sizes and interfacial area to understand the global and local effects on the process is still the optimal condition for an experimental approach. The combination of global and local measuring techniques in the complexity reduced multiphase system of Taylor bubbles enables new insights into fluid dynamics and mass transfer processes as well as the overall mass balance of the transferred species. In this chapter the experimental set-up for Taylor bubbles and the methods for the detailed investigation of global and local processes at fluid interfaces are introduced and explained [Kas15] [Kas17a] [Kas17b]. Direct optical measurement techniques, like shadowgraphy and high-speed 2D and 3D laser techniques became more reliable and highly efficient during the past decades. With these techniques an approach of combined global and local measurements is possible to enable mass balance of species transport from gas to liquid phase. This complexity reduced and well defined multiphase system can be investigated even for industrial relevant reactive systems to aid to understand the influence of timescales of mixing and chemical reaction on the yield and selectivity [Kas17b] [vK19].

## 3.1 Experimental set-up

The experimental apparatus in Fig. 3.1 consists of a vertical test section, an upper tank with a constant fill level controlled by a drainage system and either, a) high-speed cameras and LED light sources or b) two high-speed cameras and laser light sheet optics for optical measurements. The test section consists of vertical borosilicate glass channels of 300 mm in length with different cross sections. Four circular pipes, $D$ = 5.5, 6, 7 and 8 mm, and two square ducts, with a hydraulic diameter of $D_h$ = 6 and 8 mm, are used to investigate the effects of channel geometry on the transport processes. For refractive index matching, the channels are surrounded by a duct made of borosilicate glass [a) & b)]. The gap is filled with a DMSO (Dimethyl sulfoxide, Sigma Aldrich)-water solution (Fig. 3.1 right, at cross-sectional view). For optical access into the channel with minimized optical disturbance, the concentration and temperature of the DMSO solution is kept at 97 wt.-% and $T_{DMSO}$ = 298 ± 1.0 K, respectively, to obtain the same refractive index as borosilicate glass ($n_1$ = 1.473).

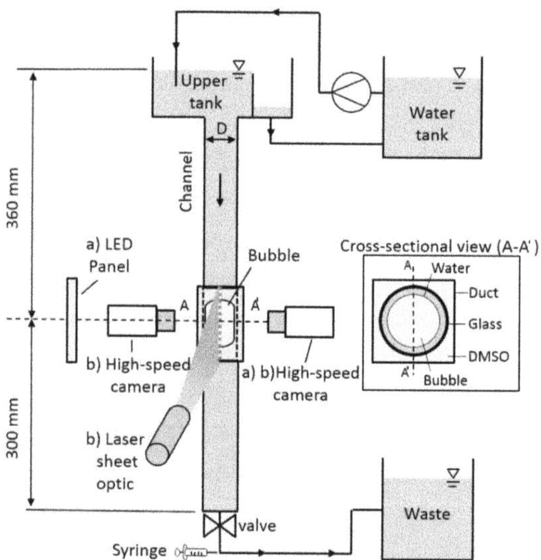


Fig. 3.1 Schematic experimental set-up after [Kas15] [Kas17a], where a) LED back-light imaging or b) laser measuring techniques with one or two opponent cameras can be used for global and local investigations of transport processes.

A predetermined amount of $CO_2$ (99.995 vol.-% purity, Westfalen AG) is stored in a precise gas-tight syringe (Hamilton®1001) and injected into the setup, filled with deionized water. The conductivity $\kappa$ of the water is less than $\kappa = 1.2\ \mu\text{S cm}^{-2}$. The liquid temperature $T_l$ is set by a thermostat at $T_l = 298 \pm 0.5$ K. To mimick the mass transfer of rising bubbles and simultaneously measure the mass transfer at the interface using the cameras the bubble is held at a constant height in a counter flow as suggested elsewhere [Abe08]. The liquid flow rate is regulated by a valve so as to keep the bubble suspended at 360 mm below the water surface in the upper tank. The constant fill level is provided by a pump and a drainage circle flow. This results in a constant hydrostatic pressure condition for all experiments at the observation area.

a) For investigations of overall mass transfer and rise velocity a high-speed camera (Optronis CamRecord 5000, frame rate: 250 fps; objective lens Sigma 20 mm F1.8 ex dg) was placed opposite to an LED Panel to visualize and record the bubble shape and its movement to calculate the bubble size and its rise velocity as explained in Chapter 3.2.1.

b) For the investigation of local fluid velocity and transferred $CO_2$ concentration a laser light source for excitation of fluorescence, a Nd:YLF laser (Darwin Duo 527-100M, $\lambda$= 527 nm, pulse width < 210 ns, pulse repetition rate 600 - 1000 Hz; Continuum) was installed. To illuminate a planar area around the fluid dynamically fixed or in stagnant liquid rising Taylor bubbles, the laser beam was widened with light sheet optics (ILA_5150 GmbH). The emitted light from the fluorescent dye (fluorescein sodium salt) or from the fluorescent tracer particles (microparticles GmbH) was recorded with a PCO Dimax HS2 (frame rate 600 - 1000 fps) placed perpendicular to the laser sheet and equipped with a band-pass filter to protect the camera chip from direct laser light (Semrock FF01-590/20, 590 $\pm$ 2 nm, half-power bandwidth 20 $\pm$ 2 nm, transmission > 84%). The methods are introduced in Chapter 3.2.2 & 3.2.3.

## 3.2 Visualization techniques & procedures

In order to measure the terminal bubble rise velocity, the local fluid velocity, the overall mass transfer coefficient of the bubbles and the local concentration in the fluid around bubbles, three different visualization techniques are used, compare Fig. 3.2 LED, PIV, LIF). LED Backlight Imaging or shadowgraphy enables to capture the bubble shape and position [Kas15], Particle Image Velocimetry (PIV) visualizes the deposition of dispersed fluorescent particles in the fluid per time [Kas17b] and Laser Induced Fluorescence (LIF) visualizes a fluorescent dye dissolved in the fluid, which indicates mass transfer into the fluid [Kas17a] [Kas17b]. The following chapter explains the imaging techniques introduced previously with the required equipment and the post processing methods, for accurate results.

Fig. 3.2 Experimental setup and exemplary images of the visualization techniques: Back-light Imaging, PIV & LIF. Left: Experimental setup with high-speed camera from the right side and the laser sheet is generated from the front. Right: Exemplary images of the three visualization techniques LED Back-light Imaging, PIV and LIF.

### 3.2.1 LED back-light imaging

LED back-light imaging (a)) enables to capture the bubble interface and shape, therefore it is used to estimate the bubble volume and its motion. The terminal bubble rise velocity in stagnant liquid is calculated by the deposition of the bubble front or the center of mass in between images with a defined time delay. The mass transfer coefficient of the bubble can be calculated by the time depending volume shrinkage of the bubble.

Therefore, images of each bubble are recorded at a fixed position in the center of the observation field of a high-speed video camera (Optronis GmbH, CamRecord5000, frame rate: 250 fps, exposure time: 1 ms, spatial resolution: 0.08 mm pixel$^{-1}$), placed right-angled to the channel and outer duct. Figure 3.3 schematically shows the image processing method for instantaneous bubble volume and diameter measurements. An original bubble image (Fig. 3.3 (a)) was transformed into a binary image (Fig. 3.3(b)). Since the refractive index of water

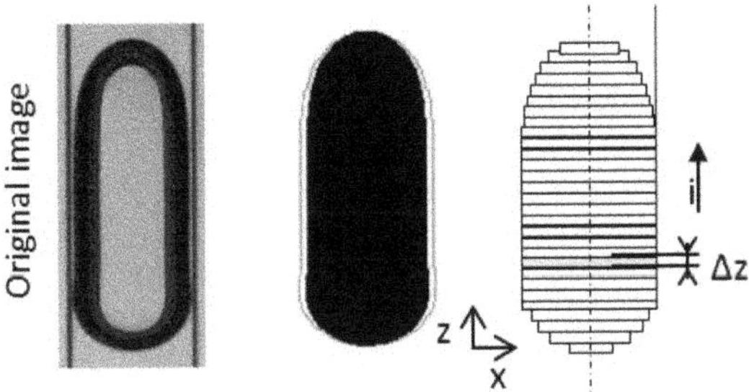


Fig. 3.3 Post processing method to calculate the sphere-volume equivalent bubble diameter $d_{eq}$ from 2D backlight images. a) original image; b) binary image and red corrected bubble contour; c) disk image with the radius of the corrected contour [Kas15].

($n_1 = 1.333$) flowing inside the channel differs from the one of the channel glass ($n_2 = 1.473$) the angle of refraction differs to the angle of incident light hitting the camera. This distortion is caused by the change of speed of light at the interface of two media with different optical densities. Therefore, the camera observed a virtual projection of the bubble with smaller dimensions, where the real bubble would have been slightly underestimated. The binary image was corrected using the Snell's law, where the real bubble interface position is shown with the outer line (Fig. 3.3(b), compare Fig. 3.4).

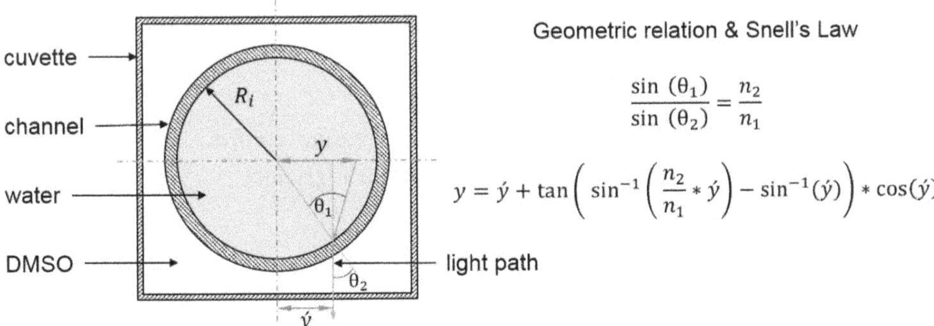


Fig. 3.4 Refractive index matching and correction via Snell's Law. Geometric relation in the channel cross section and the distortion corrected by Snell's law, due to the difference in refractive index inside the channel.

The method assumes that all the horizontal cross sections of a bubble are circular, which is applicable for low $Eo$ systems with less dynamic shape oscillation or interfacial waves. The height $\Delta z$ of each circular disk is a physical length of one pixel equivalent. The resultant circular disks are piled up in the vertical direction to reconstruct a three-dimensional bubble shape (Fig. 3.3 (c)).

The sphere-volume equivalent bubble diameter $d_{eq}$ is evaluated using

$$d_{eq} = \sum_{i=1}^{N} (6{r_i}^2\,\Delta z)^{1/3} \tag{3.1}$$

where $N$ is the total number of pixels in the vertical direction, $i$ the index denoting the pixel number in the vertical direction and $r$ disk radius. This post processing method is verified by analyzing images of a rigid sphere of 5 mm in diameter falling through $D_h = 6$ mm. The uncertainty estimated at 95% confidence in equivalent diameter $d_{eq}$ determined from the measurement of a rigid sphere is less than 1%. Accurate equivalent diameters $d_{eq}$ for bubbles in square ducts, however, cannot be obtained using this method as is, since their horizontal cross-sections were not axisymmetric [Bod14] [Hag16]. The linear regression curves, $d_{eq} = a \cdot d_{eq*} + b$, where $d_{eq*}$ is the measured apparent equivalent diameter, for the bubbles in the square ducts were, therefore, developed by comparisons between injected and measured $d_{eq}$ of air bubbles as shown in Fig. 3.5. The parameters $a$ and $b$ were estimated to be 1.10 and $-0.528$ for $D_h = 6$ mm, and 1.09 and $-0.593$ for $D_h = 8$ mm, respectively. The rise velocities $v_B$ of bubbles in stagnant liquid in the channels are measured using the vertical position $z(t)$ of a bubble nose where $t$ is the time. The $z$ data are fitted with a linear function using the least square method to obtain $v_B$. Uncertainty estimated at 95 % confidence in $v_B$ is 0.05 %.

### Evaluation of global mass transfer coefficient

The mass transfer coefficient $k_L$ is evaluated from the decreasing rate of $d_{eq}$ [Hos14a] [Kas15]. The rate of change in the mole $M$ of $CO_2$ inside a bubble is given by

$$\frac{dM}{dt} = -k_L\,A\,(C_c^* - C_c), \tag{3.2}$$

where $A\ (= \pi\, d_{eq}^2)$ is the bubble surface area calculated with the equivalent bubble diameter $d_{eq}$, $C_c^*$ the $CO_2$ concentration at the bubble interface (34 mol·m$^{-3}$) and $C_V$ the $CO_2$ concentration in water. The $C_c^*$ is derived by Henry's law, which is given by

$$pX = \frac{C_c^*}{C_c^* + C_V} H, \tag{3.3}$$

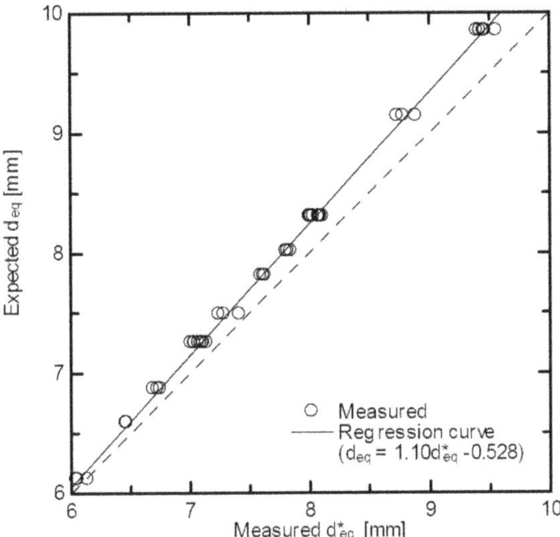


Fig. 3.5 Regression between $d_{eq}$ and $d_{eq*}$ in a square duct $D_h$ = 6 mm evaluated by air bubbles [Kas15]

where $p$ is the pressure inside the bubble, $X$ the mole fraction of $CO_2$, $C_V$ the molar concentration of water (55.4 kmol·m$^{-3}$) and $H$ the Henry constant (166 MPa for $CO_2$ at $T_L = 298$ K). Since the dissolution rates were very small in all the experimental runs, $X$ can be postulated to be unity. $C_V$ is much larger than $C_c^*$, i.e. $C_V/C_c^* \approx 1600$, so that $C_c^*$ is given by

$$C_c^* = \frac{C_V p}{H}, \tag{3.4}$$

The pressure p is given by

$$p = p_{atm.} + \rho_L \, g \, z, \tag{3.5}$$

where $p_{atm}$ is the atmospheric pressure and the additional acting hydrostatic pressure by water column above the bubble nose, i.e. $z = 360$ mm with the density $\rho_L$. Eq. 3.5 presumes that the contribution of surface tension is negligible. Substituting Eq. 3.4 into Eq. 3.2 and neglecting $C$ yields

$$k_L = \frac{1}{\pi \, d_{eq}^2} \frac{H}{C_V p} \frac{dM}{dt}, \tag{3.6}$$

By assuming that $CO_2$ is an ideal gas, we can express $dM/dt$ in terms of $p$ and $d_{eq}$ as follows:

$$\frac{dM}{dt} = \frac{\pi}{6 \, R_m \, T_L} \frac{d(p \, d_{eq}^3)}{dt}, \tag{3.7}$$

where $R_m$ is the universal gas constant. Substituting Eq.3.7 into Eq.3.6 and evaluating $d(pd_{eq}^3)/dt$ by using a centered difference scheme yields

$$k_L = \frac{H(p_2\, d_{eq,2}^3 - p_1\, d_{eq,1}^3)}{6\, R_m\, T_L\, C_V\, p_{12}\, d_{eq,12}^2(t_2 - t_1)}, \tag{3.8}$$

where the subscripts, 1, 2 and 12, represent the time $t_1$, $t_2$ and $t_{12} = (t_1 + t_2)/2$, respectively. Since $p$ was constant in the experiments, Eq. 3.8 becomes

$$k_L = \frac{H(d_{eq,2}^3 - d_{eq,1}^3)}{6\, R_m\, T_L\, C_V\, d_{eq,12}^2(t_2 - t_1)}, \tag{3.9}$$

The bubble diameters, $d_{eq,1}$ and $d_{eq,2}$, at $t_1$ and $t_2$ were obtained from a linear regression for measured time evolution of $d_{eq}$ as shown in Fig. 3.6. The bubble shape change is also shown in the figure. All the data of $d_{eq}$ are well fitted by linear functions. The $d_{eq,12}$ is calculated as $d_{eq,12} = (d_{eq,1} + d_{eq,1})/2$.

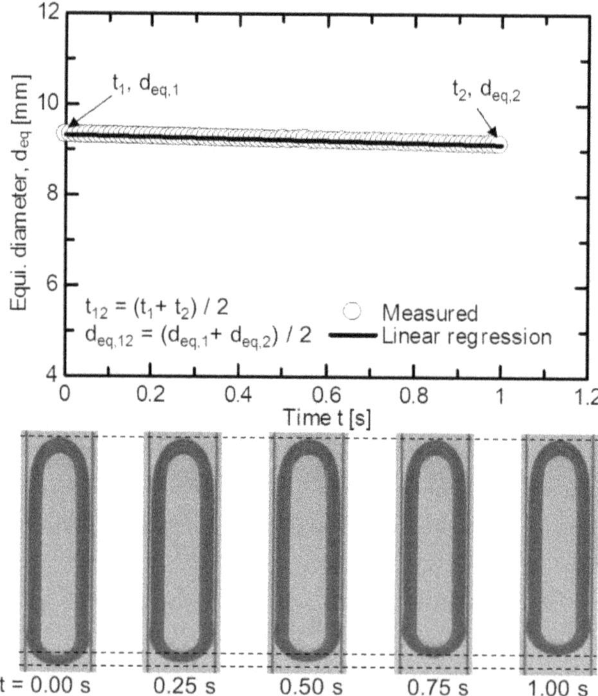


Fig. 3.6 Example of measured time evolution of $d_{eq}$ during the dissolution of a $CO_2$ Taylor bubble in $D$ = 6 mm channel [Kas15]

### 3.2.2 Particle Image Velocimetry - Evaluation of velocity fields

Accurate local information in time and space of fluid dynamics can be evaluated by using the non-invasive measuring technique Particle Image Velocimetry (PIV). The most common technique is to evaluate inside a quasi two dimensional laser light sheet (2D) two components (2C) of the flow velocity (2D2C-PIV). Therefore, optical access to the system and buoyancy neutral tracer particles are required. Those particles follow the fluid motion and their displacement ($\Delta s_x$, $\Delta s_y$) in time ($\Delta t$), can be evaluated by cross-correlation, where the resulting vector field can be assumed to represent the velocity field of the fluid motion. The main equipment for PIV are usually synchronized pulsed laser with light sheet optics and a perpendicular a camera recording the displacement of the particles in the fluid motion, as shown in Fig. 3.7.

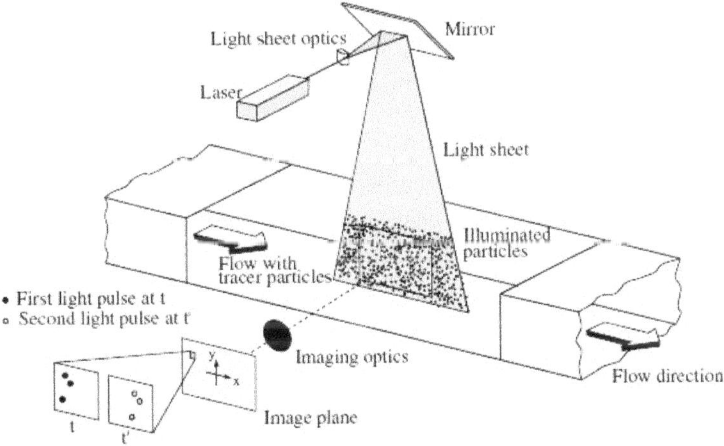


Fig. 3.7 Scheme of a typical experimental set-up of PIV measurement in fluid flows [Raf18]

The fluid dynamical problem, which is investigated, dictates the required Field of View and magnification for the spatial resolution and the maximal time difference between images for the temporal resolution. In the following, the sophisticated laser measuring technique PIV will be roughly introduced. It is a complex research field by its own and for further interest the literature from Adrian [Adr11] and Raffel [Raf18] is highly recommended. Buoyancy neutral tracer particles are added to the continuous fluid phase, which are much smaller than the smallest flow structure to be analyzed (e.g. vortices) but large enough to cause a bright response on the camera chip. Those tracer particles are often coated with a fluorescence dye, which is excited by the 2D laser light sheet (e.g. $\lambda_{ex.} = 527$ nm) and emits a longer wavelength (e.g. $\lambda_{em.} = 590$ nm). The advantage of wavelength conversion instead of direct light scatter, is the option to protect the camera chip from high energetic light via optical filters and the improvement of image quality by reduction of reflections, e.g. at interfaces. A pulsed laser and a cylindrical lens systems is used to generate thin illumination planes inside the experimental volume. E.g. Nd:YLF lasers (Neodym-doped Yttrium Lithium Fluoride) enable high energetic pulses with a repetition rate of several kHz and a wavelength of 527nm. Another advantage of this laser is the equidistant time series of pulses ($\Delta t_{i,i+1} = const.$), which leads to better accuracy and new insights into dynamic changes in the velocity field. Modern high-speed cameras (CMOS - Complementary Metal Oxide Semiconductor) are required to achieve high frame rate recordings. A frequency generator or synchronizer, which is used to trigger both laser and camera for perfectly exposed images. As it is illustrated in Fig. 3.7, two consecutive images are used to calculate the velocity on the basis of the known temporal distance between the images $\Delta t$, the magnification factor Q, and the displacement of the tracer particles $\Delta s$:

$$\vec{v} = (u_x, v_y) = \frac{(\Delta s_x, \Delta s_y)}{Q \cdot \Delta t_{i,\, i+1}}, \tag{3.10}$$

The magnification factor [mm/pix] is usually calculated using calibration images of targets with defined patterns of lines and dots, which are placed at the observation area inside the experimental setup. An alternative is to use defined dimension components inside the recorded images, like needles, spheres or wall thicknesses or distances. With these information a vector field can be calculated, where PIV detects the displacement of particle groups inside defined interrogation areas (windows), by space-time cross-correlation. The advantage of the analysis of particle groups instead of single particles, is a higher probability to assign a displacement of a characteristic pattern of peaks in a double image compared to single peaks. Typical sizes of these areas are 64x64 to 16x16 pix$^2$, where this area should picture minimal 5 tracer particles, but the optimal window size depends also on the particle size, particle displacement and the particle density inside fluid. Since 1995, PIV has been applied in a broad field of multiphase flows and is nowadays a state of art measuring technique [Adr11].

To determine the fluid flow structure around Taylor bubbles rising in stagnant liquid or being hydrodynamically fixed in counter current flow, the following equipment and settings have been used: A solution of tracer particles (microParticles GmbH, PS-FluoRot-1.5 or PS-FluoRot-3.0 with $d_p = 1.56$ or $3.16\mu$m, respectively) was added to the liquid. As a light source for excitation a Nd:YLF laser (Darwin Duo 527-100M, $\lambda = 527$ nm, pulse width $< 210$ ns, pulse repetition rate 600 Hz, Continuum) was installed. To illuminate a planar area around the hydrodynamically fixed or rising Taylor bubbles, the laser beam was widened with light sheet optics (ILA5150 GmbH). The light from the tracer particles was recorded with a PCO Dimax HS2 (600 - 1000 fps) placed perpendicular to the laser sheet with equipped bandpass filter to protect the camera from direct laser light (Semrock FF01-590/20, 590 $\pm$ 2 nm, half-power bandwidth 20 $\pm$ 2 nm, transmission $> 84$ % ). For PIV post processing the commercial software "PIVview2c V3.62" (PIVTEC GmbH) was used. An overview of the experimental and post processing settings is given in the Appendix C in Table C.1. [Kas17b]

For the presentation of the correlated velocity fields, coordinates as seen in Fig. 3.8 are used. The $z*$ coordinate is used here as relative coordinate, where the bubble tip always represents $z^* = 0$ mm for a better understanding of the plotted velocity fields and velocity profiles.

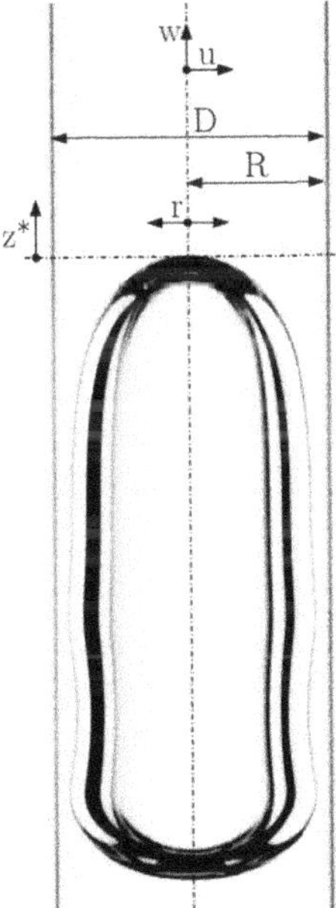


Fig. 3.8 Local coordinates for 2D p-LIF or 2D2C PIV measurements, exemplary at a Taylor bubble in $D = 7$ mm.

In Fig.3.9 the scheme of the 2D2C PIV measurement in parallel planes in a square channel is shown. With this technique the consecutive evaluation of the flow conditions in decentered planes of a non rotation-symmetric flow is enabled, if mass transfer can be neglected and the bubble shape is stationary.

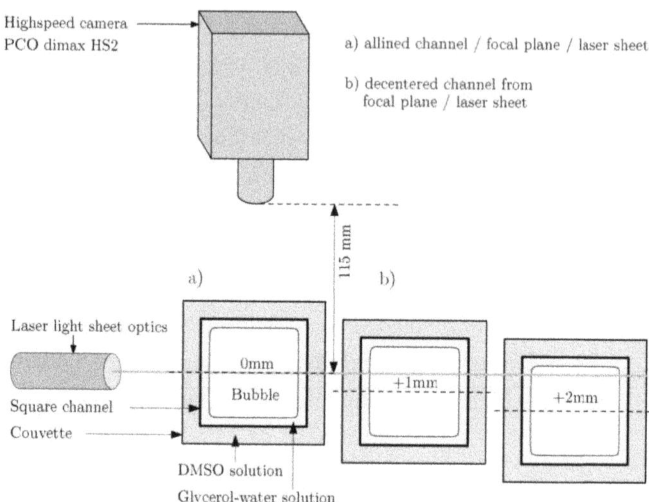


Fig. 3.9 Schematic cross sectional view of PIV measurements in different planes within a square channel setup $D_h$= 6 mm. a) Channel center aligned with focal plane and laser light sheet; b) Different decentered channel positions.

### 3.2.3 Laser Induced Fluorescence - Evaluation of concentration fields

#### Planar 2D Laser Induced Fluorescence - p-2D LIF

For the local investigation of mass transfer processes another indirect fluorescence based measuring technique is needed, as introduced before. Laser Induced Fluorescence is based on the addition of a fluorescence dye to a liquid phase, where a laser excites the dye and the emitted light is recorded by a camera. Dyes are considered to act as a passive scalar and to be non-invasive like tracer particles for PIV. The fluorescence intensity depends on various factors as the dye concentration, excitation energy and wavelength and the presence of possible quencher, which can be a transferred species or a fluid property. Therefore, LIF is used for two main applications in two-phase flows: the direct measurement of local dye concentration or an indirect measurement of a quencher concentration in the liquid. If a dye is dissolved into a liquid phase and mixed in a binary mixer with another liquid phase without any dye, the fluorescence intensity gradient directly indicates the mixing state, due to the local dilution of the dye concentration. This direct approach of LIF had a high impact on the investigations of laminar mixing structures and mixing rates in microchannels [Hof06]. The indirect approach establishes the measurement of a local quencher concentration, where the excited dye transfers the higher energy state to the quencher and the radiation of electromagnetic light is inhibited. This means, that a dye can be sensitive for a certain species e.g. $O_2$ or $CO_2$ and is therefore suitable for mass transfer measurements across interfaces [Bor05] [Fra11] [Die13] [Sai15] [Tim18].

For the approach of the indirect measurement of the fluorescence quencher quantity, usually the Stern-Volmer equation is used to fit a calibration curve:

$$\frac{I_O}{I_Q} = 1 + K_{SV} \cdot C_Q, \tag{3.11}$$

where $\frac{I_O}{I_Q}$ is the ratio of fluorescence intensity in absence of the quencher divided by the fluorescence intensity in presence of the quencher, $K_{SV}$ is named Stern-Volmer constant for non-dynamic quenching, and $C_Q$ is the concentration of the quencher. The excitation wavelength should be in accordance with the maximum absorption wavelength of the dye for a high yield, but can be compensated by higher excitation energy. [1] The emitted light is recorded with a camera equipped with a band-pass filter adapted to the emitted wavelength spectrum of the dye. By recording a series of reference images with presetted quencher concentrations in the liquid phase, calibration curves can be fitted for every pixel in the image.

This work uses the sensitivity of a fluorescent dye to a transferred species or a fluid parameter, and the derived fluorescence intensity indicates the value of e.g. $CO_2$ concentration in a fluid per image pixel. More precisely, here the specie is $CO_2$ and the corresponding dye is fluorescein sodium salt (also known as "Uranin" or "Acid Yelow 73", $\lambda_{ex/em} = 460/515$nm, F6377 Sigma Aldrich/Merck). The excitation of the fluorescence dye is accomplished with a $\lambda$=527 nm diode-pumped Nd:YLF laser, Darwin-Duo 527-100-M from Quantronix, where laser sheet optics creates a planar laser sheet inside the observation volume [2]. Perpendicular to the laser sheet, high-speed video cameras are integrated to capture the fluorescence intensity of the fluorescein around the bubbles. Two high-speed cameras can be used to observe the multiphase flow with two different spatial resolutions, where a local view at the interface needs to be orientated inside the global view, which shows the Taylor bubble and wake. The experimental setup is basically the same then as is used for PIV.

[1] To find the optimal fluorescence dye for an application, the database fluorophores.org from the TU Graz is recommended. The database can be browsed by excitation or emitting wavelength, solvents or even common applications. Some entries are including measured excitation and emitting spectra and some publication references are given, where the dye has been used.

[2] The excitation wavelength of the laser is far away from the optimal absorption maximum of fluorescein, but the lower intensity of the dyes absorption spectra can be compensated with a higher excitation energy of the laser pulse. Compare with the Appendix B in Fig. B.1 showing the exemplary plot of laser excitation wavelength, absorption and emission bands of fluorescein and the filter transmission band.

During all measurements and the calibration procedure the fluid properties like the conductivity $\kappa$, the pH value (WTW MultiLine ® 3510 IDS, TetraCon® 925, SenTix® HW-T 940, Xylem analytics), the $CO_2$ concentration $C_{CO2}$ (pCO2mini with sensor spots, PreSens GmbH) and the temperature $T_l$ are measured inside the upper tank, before the fluorescein solutions is flowing down through the test section. In Fig.3.10 the time evolution of these parameters are plotted over time during the procedure of taking calibration images. In the beginning at $t = 0$ s the deionized water dye solution is in equilibrium with air. The conductivity is very low, the $CO_2$ concentration or here partial pressure is nearly zero and the pH value takes some time approaching a constant value, which is caused by the very low ions concentration. After the $CO_2$ gassing is started $t > 1000$ s, $\kappa$ and $C_{CO2}$ is increasing directly and reaches saturation in less than 180s, while the pH value drops. This is caused by the dissociation of the $CO_2$ in water. The reference images are taken for a $CO_2$ saturated state. At $t = 2000s$ the gassing with $CO_2$ is stopped and the fluid is pumped through the experimental setup, where the $CO_2$ concentration is decreasing slowly due to mass transfer into the ambient air in the open setup. The conductivity is decreasing in parallel with the dissolved $CO_2$, while the pH value increases again.

During the degassing process, the reference images were taking every 150 s for about 3500 s until the initial state at $t = 0$ s (air equilibrium) is nearly reached, which ends up in about 20 data points and calibration image sets for the evaluation of calibration functions. An exemplary averaged calibration of dissolved $CO_2$ versus the grayscale values in the images is shown Fig. 3.11, where the mean values of dissolved $CO_2$ are plotted over mean grayscale values. These three values are measured simultaneously during the calibration. With such calibration functions a direct calculation of the local dissolved $CO_2$ or local pH-Value can be made from the grayscale of the image pixels, which stands for the fluorescence intensity. The fitting of the points is best with a two-term exponential function, which is calculated for every pixel of the camera chip. This pixelwise procedure takes into account heterogeneous light excitation in images. This is important for LIF, because every light sheet has a heterogeneous intensity due to the laser beam intensity distribution, beam quality

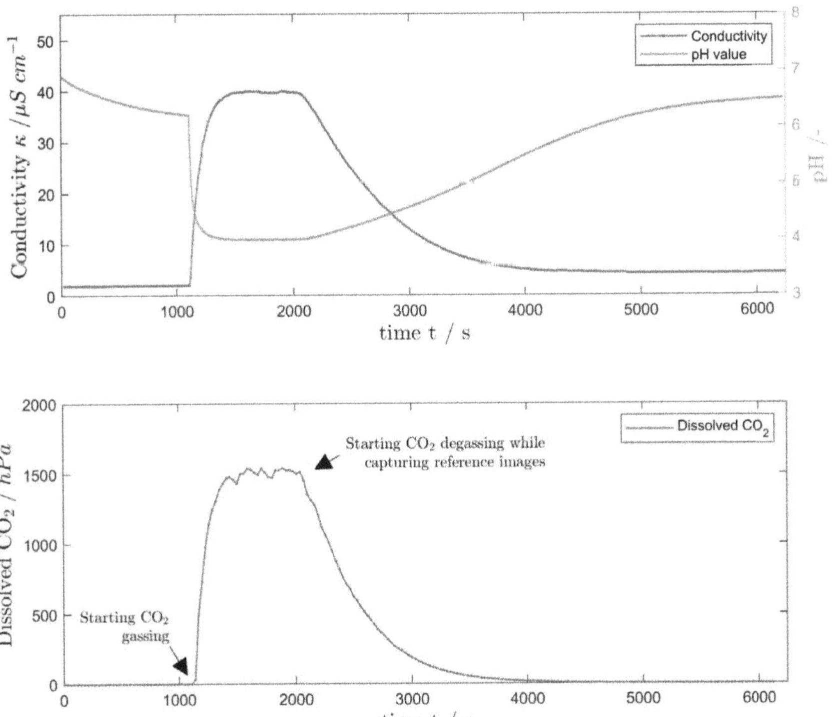


Fig. 3.10 Measured fluid properties like pH, conductivity and dissolved $CO_2$ while capturing representative images. This enables pixelwise calibration for the estimation of local dissolved $CO_2$ concentration fields for LIF.

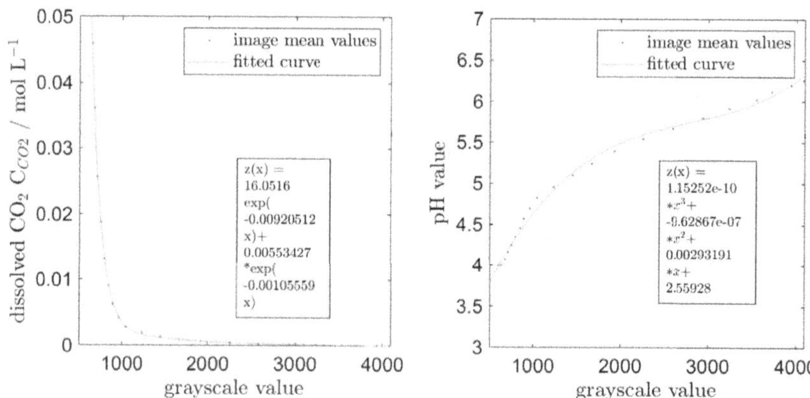


Fig. 3.11 Schematic LIF calibration function of dissolved $CO_2$ concentration and pH value vs. the mean image grayscale. The actual method generates a matrix of pixelwise calibration functions to take into account the local excitation for the exact dissolved $CO_2$ estimation as well as sensitivities of the individual camera-pixels.

and due to light extinction along the light path through the fluid phase. Furthermore, also differences in pixel sensitivities are accounted for.

The pixel specific calibration functions are used to calculate the resulting local $CO_2$ concentration in the equivalent 'fluid element', which is represented by a pixel in the image. This enables detailed $CO_2$ concentration fields at Taylor bubbles in channels recorded with high frame rates to understand and investigate even dynamic processes of vortex shedding or stretching of concentration peaks due to wall shear flow. Even long-term dissolution processes of Taylor bubbles in counter current flow with a development of the concentration field in magnitude and phenomenology can be observed. Laser induced fluorescence is an ideal optical technique for the visualization and quantification of local transport processes in two-phase flows.

Note that many dyes' fluorescence intensities are depending also on temperature, which can be used as indirect temperature measurement as well, but otherwise illustrates the importance to operate experiments under constant conditions. Therefore, the pulse energy of the exciting lasers should not heat up the fluids or solid walls.

### Mass Balance between Local and Global Mass Transfer

The well defined conditions and rotationally symmetry in this unique two-phase flow system of Taylor bubbles is optimal for a mass balance approach using global and local measuring techniques for mass transfer processes. The approach for a mass balance is the validation of the volume reduction $\Delta V$ (Eq. 3.12) of a rising $CO_2$ Taylor bubble in stagnant liquid between $t_1$ and $t_2$ (compare Fig. 3.12) and the integral of the $CO_2$ concentration field $C_{CO2}$, which has been established during the same $\Delta t$. In Fig. 3.12 the approach of the mass balance is sketched, where the volume reduction and the rotation symmetric wake structure (dark area) is clearly visible.

$$V_{1,2} = \sum_{i=1}^{n} (\frac{\pi\, d_i{}^2\, \Delta z}{4}), \tag{3.12}$$

Taking into account the ideal gas law, the moles of the $CO_2$ Taylor bubble at $t_1$ and $t_2$ can be calculated as

$$M_1 = \frac{p_1(\pi\, d_1{}^3/6)}{R_m\, T}; \ M_2 = \frac{p_2(\pi\, d_2{}^3/6)}{R_m\, T} \tag{3.13}$$

and the difference in moles at the two timesteps, should be equal with the integration of the rotated $CO_2$ concentration field measured at time $t_2$, which is indicated by the green region in Fig. 3.12. This assumption can be written as Eq. 3.14

$$M_1 \quad M_2 = B \cdot 2\pi \int_{r=0}^{r=1} \int_{z=z_1}^{z=z_2} C_{CO2}(r,z) r\, dr\, dz \tag{3.14}$$

and is actually a 3D reconstruction of dissolved $CO_2$ concentration wake from 2D p-LIF image. This method requires obviously a flow structure and a concentration field, which is rotationally symmetric. Otherwise, the integration would loose its eligibility, therefore Taylor bubbles in channel $D = 6$ mm and $D = 7$ mm with neglectable bubble shape dynamics and a nearly steady fluid flow structure in the wake are investigated with this approach.

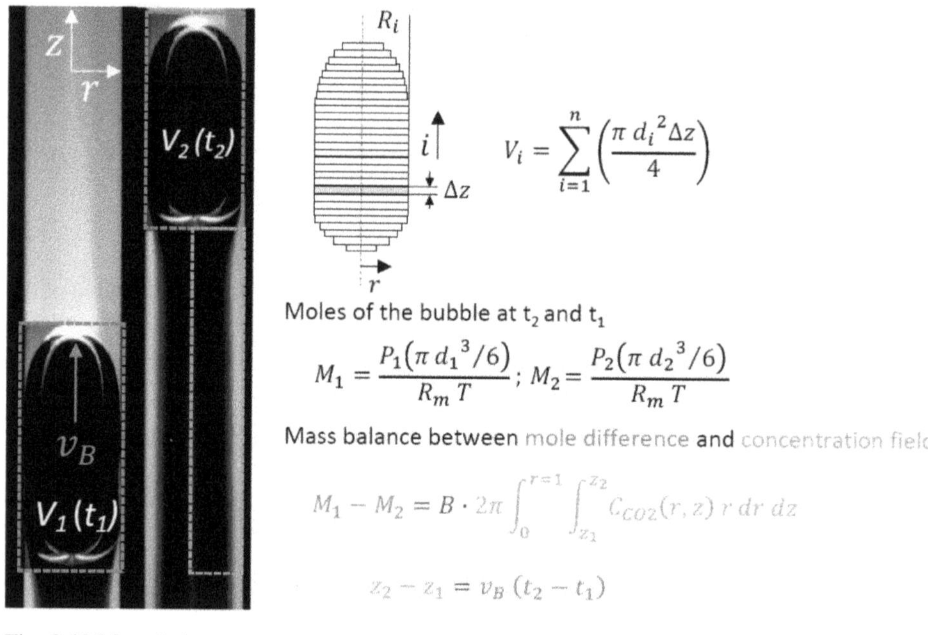


Fig. 3.12 Mass balance approach for the validation of global and local mass transfer at a $CO_2$ Taylor bubble e.g. in $D = 6$ mm channel ($Eo = 4.9$, $Re = 36$). $CO_2$ mole difference of the bubble volumes at $t_2$ and $t_1$ is compared with the integrated $CO_2$ moles in the concentration field (green region).

## 3.3 Procedures and experiments

### 3.3.1 Preparation of the experimental setup and working fluids

The conscientious preparations for all different experiments have been very similar, whether quantitative analysis of the rise velocity, mass transfer coefficient, velocity or concentration fields at wake structures behind Taylor bubbles were realized. The experimental setup is cleaned with at least 10 Liters of deionized water, to dilute and wash out possible contamination from dust, even if the open setup parts are covered by wrapping film while not in use. All tubes were exchanged to reduce the risk of cross contamination from tracer particles or fluorescent dyes, to reduce disturbance for the following optical method. Even though, the tracer and dye are non-invasive for the flow, they can interact with each other, e.g. remaining tracer particles in the system prohibit to measure a local $CO_2$ concentration at their position, due to their own fluorescence. In case of experiments in contaminated systems by usage of the Surfactants Triton X-100, the experimental setup was washed with hot water for 1h, before changed over to deionzed water for 1h. After the intensive washing procedure, all tubes were exchanged, before the washing was continued with deionzed water to lower the risk of cross contamination from the new tubes. During the washing procedure and all experiments the conductivity of the liquid phase was measured and tracked at the top of the channel to ensure ideal conditions and the operational reliability of the deionization process.

The liquid phase for all experiments is based on deionized water from an in-house facility with a very low conductivity of $\kappa(T = 298\ K)=1.3\ \mu S\ cm^{-1}$, where the laboratory tap water has a mean conductivity of $\kappa = 280\ \mu S\ cm^{-1}$. Depending on purpose, additional tracer particles, dye or Surfactants are added to prepare the working solution, which is saturated with air. PIV measurements required the addition of 30 $\mu L\ L^{-1}$ of a 2.5 wt% solution of solid polystyrene tracer particles (PS-FluoRed-1.5/Fi189; $d_p = 1.5 \mu m$; $SD = 0.04\ \mu m$; $\lambda_{abs/em} = 530/607$ nm; microParticles GmbH). The addition of the tracer particles showed no measurable effect on liquid properties as surface tension $\sigma$, viscosity $\nu$ or density $\rho$.
For all experiments the temperature of the liquid phase is set to $T_L = 298 \pm 1$ K by a thermostat for continuity and to reduce temperature effects on solubility, surface tension or viscosity. For LIF measurements the fluorescence dye fluorescein sodium salt (also known as "Uranin" or "Acid Yellow 73", F6377 Sigma Aldrich/Merck) is added in preparation of a 10 mmol $L^{-1}$ solution. In case of using laser equipment, a stable laser pulse energy has to be ensured by warming up the system before experiments to lower intensity fluctuations.
For detailed flow analysis within several parallel planes, a stable air bubble is needed to be kept in counter current flow for an unusual long time period of about 20 minutes. Therefore,

mass transfer had to be neglected by choosing air bubbles. The only bubble state, where no shape oscillation occurred and the bubble shape stayed stable until all four parallel planes have been measured, was a small bubble in a $D_h = 6$ mm channel filled with an aqueous solution of 58 wt% glycerol ($\nu$ =8 mPa s). Of course, the flow conditions within a fluid of this comparable high viscosity is very different to the flow structures in water, but the unusual phenomena captured in this experiment make up for this deviation.

The accuracy of the injection volume was measured by comparing the injected air volume with the recorded and reconstructed bubble volume. Before injection of a new bubble volume, the valve is opened completely for a few seconds to ensure the channel is filled with fresh liquid and all possible remaining gas inside the channel is washed out to minimize the probability of bubble coalescence and therefore the reduction of $CO_2$ mole fraction inside the bubble. Additionally, all measuring equipment is checked and ready to record the bubble entering the Field of View of the cameras. After the recording, the valve is opened completely to wash out remains of the gas bubble and to ensure no gas volume enters the upper tank or remains within the setup. The gaseous phase for the bubbles was changed in function of to the application. Ambient air was used for bubble volume calibration images, the determination of bubble rise velocities or some PIV measurements, because the liquid phase is in equilibrium with the ambient air, which is ideal for the measurement of hydrodynamics decoupled from mass transfer processes. On the other hand, $CO_2$ was used for the investigation of mass transfer processes. The $CO_2$ has a purity of 99.995 vol % (Westfalen AG) and is supplied by gas bottle directly at the experimental setup via a T-junction tubing and a septum. Here the gaseous phase is withdraw with a gas-tight 1mL glass syringe (Hamilton® 1001), with small injection needle ($d_{in} = 0.5$ mm, Sterican®, B. Braun Melsungen AG), after three full fillings the syringe with $CO_2$ and subsequent disposal. The last full filling of the syringe is reduced to the exact gas volume for injection directly in front of the septum of the experimental setup (Fig.3.1), to lower the chance of counter diffusion of ambient air into the syringe.

### 3.3.2 Experiments

During this work, many experiments have been carried out to investigate different aspects regarding the transport processes at Taylor bubbles in vertical channels. The characterization of bubble flow regimes is more accurate using air bubbles, which are in equilibrium with the deionized water, rising in stagnant fluid in various channels with different diameters $D = 5.5 - 8$ mm and cross sections (circular & square) or kept in counter current flow (CCF). Therefore, the mass transfer processes in these experiments are negligible for a fluid dynamic characterization. $CO_2$ is used to investigate the mass transfer processes at Taylor bubbles in the same channels to extend the range of existing mass transfer correlations to Taylor bubbles in millimetric scale channels. Local investigations of the fluid motion around the bubbles help to identify characteristic bubble wake structures depending on the channel diameter $D$ or on the rise velocity and therefore rising regimes of bubbles. Local mass transfer investigations via visualization of local $CO_2$ concentration fields enable to identify the concentration boundary layers at the bubble interface and the mixing within the wake and the bulk of the fluid phase. Additionally, the combination of global and local mass transfer measurements can provide a mass balance of the transported gaseous phase into the liquid phase. In the following list, the rough structure of the experimental results is summarized:

- Rise velocity of Air and $CO_2$ small and Taylor bubbles in clean and contaminated water (CCF, stagnant) / circular and square in Chapter 4.1
- Local velocity field at Taylor bubbles via high-speed PIV (CCF, stagnant) / circular and square channels in Chapter 4.1
- Global mass transfer of $CO_2$ small and Taylor bubbles in clean and contaminated water (CCF, stagnant) in Chapter 4.2
- Local concentration fields at $CO_2$ Taylor bubbles via high-speed LIF (CCF, stagnant), to get wake structures and mixing rates in Chapter 4.2
- Mass balance between the volume reduction of the bubble and the local $CO_2$ field behind Taylor bubbles in Chapter 4.3

# Chapter 4

# Results and discussion

To clarify the inter-dependencies of local and global fluid dynamics and mass transfer processes, different experimental methods are used to investigate air and $CO_2$ bubbles in vertical channels. The results of these investigations are structured according to their affinity and level of complexity and discussion of the corresponding results is immediately attached to each subsection.

The investigations show, that the wall dominant regime of Taylor bubbles leads to reproducible and adjustable rise velocities, mass transfer coefficients and wake structures, by adjusting the channel diameter only. Additionally, detailed studies of the local mass transfer processes enable a mass balance from $CO_2$ Taylor bubbles rising in vertical channels. Partly, this research has been carried out within the priority program DFG SPP1506[1] 'Transport Processes at Fluidic Interphases' of the German Research Foundation. Thus, Taylor bubble experiments enable detailed investigations of the influence of wake structures and local mixing on chemical reactions.

Therefore, Taylor bubbles have become one of the guidance experiments of the priority program DFG SPP1740[2] of the German Research Foundation. At the end of this chapter ongoing research on reactive Taylor bubbles will be introduced briefly.

[1] http://www.dfg-spp1506.de/
[2] http://www.dfg-spp1740.de/

## 4.1 Investigation of global and local fluid dynamics of bubbles in channels

### 4.1.1 Flow regimes of bubbles rising in channels - Classification in small, intermediate and Taylor bubbles

The global fluid dynamics of bubbles in vertical channels have been investigated for more than a century by various researchers who measured the terminal rise velocity of bubbles in channels with different sizes and shapes [Dum43] [Tay61] [Whi62] [Bro65]. They all agree, that elongated bubbles or Taylor bubbles have a rise velocity $v_B$, which is independent of their bubble volume as long as the diameter ratio $\lambda = d_{eq}/D$ is larger than a critical value $\lambda \approx 1$ [Hos14a] [Hay14]. The location of the critical value $\lambda$ can be estimated by the crossing point of the two linear regressions of the two regimes of small and Taylor bubbles ($Eo > 10$) (compare Fig. 2.12) [Mae15] [Kur13]. Usually, the first regression stands for the rise velocity of small bubbles in channels ($\lambda < 1$), which is mainly decreasing with increasing equivalent bubble diameter like in Fig. 4.1 and the second regression stands for the rise velocity of the Taylor bubbles ($\lambda > 1$) , which is independent of equivalent bubble diameter (compare Fig. 2.12). In this Taylor regime the forces are balanced and either the buoyancy force or the viscous force are depending linearly on the bubble length, which results into a rise velocity independent of the bubble volume. The hypothesis is, that the fluid film thickness becomes more or less constant, which depends on the fluid system and the channel geometry, which leads to many different empirical critical $\lambda$ values for each experimental data set. Even though, many researchers found a small transition regime in their data, where the terminal rise velocity is lowest, shortly before the bubbles are Taylor bubbles, there is no proven explanation of the transition regime from small to Taylor bubbles. The reason is, that in larger channels ($Eo > 10$), small interfacial waves and small bubble deformation influencing the fluid film between bubble and wall. Therefore, there is no steady state, which generates a reliable dependency between e.g. film thickness and rise velocity. Although, bubbles in small channels ($Eo \approx 4$) do not show interfacial waves but the fluid film thickness becomes very small and difficult to access with optical imaging measurements systems. The interfacial tension becomes more dominant, which allows only for very small rise velocities of bubbles in vertical channels, which adds to finding reliable dependencies.

In this work, the focus is on Taylor bubbles with moderate interfacial waves, because stable conditions and a steady shape of the interface are needed to enable accurate reproducible results. Due to the higher influence of the interfacial tension for low Eötvös numbers,

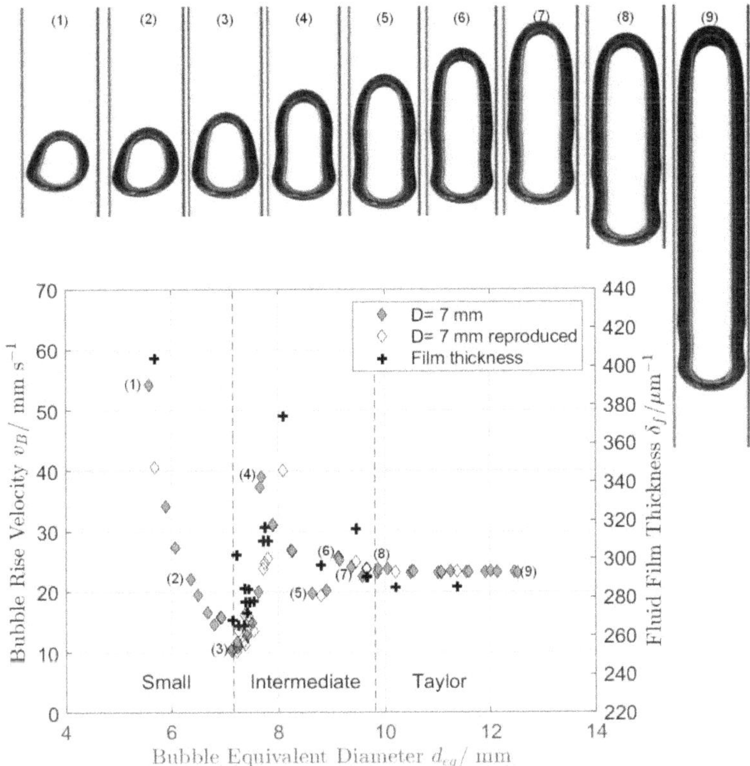


Fig. 4.1 Flow regimes of rising bubbles in $D$ = 7 mm channel, where the rise velocity $v_B$ caused by size, shape, interface curvature of the bubble and the resulting minimal fluid film thickness $\delta_f$.

$Eo = g\, \Delta\rho\, D^2\ /\ \sigma$ and a range of small channels $D = 5.5 - 8$ mm. Due to the small step variation of the channel diameter, while keeping the interfacial tension constant, it is more likely to find new insights into the dependencies of fluid film thickness and the bubble rise velocity.

Especially, bubbles rising in $D = 7$ mm ($Eo = 6.7$) channel show an extraordinary rise behavior with increasing bubble diameter (filled diamonds), where the rise velocity shows a pronounced minima ($d_{eq} \approx 7.2$ mm, $v_B \approx 10$ mm s$^{-1}$) and maxima ($d_{eq} \approx 7.8$ mm, $v_B \approx 40$ mm s$^{-1}$) with increasing $d_{eq}$, before the constant rise velocity regime of the Taylor bubble is reached ($v_B \approx 22$ mm s$^{-1}$). Because of this unique slope shape, these bubbles are analyzed in more detail, to get a deeper understanding into the transition regime and to figure out why Taylor bubbles have a volume independent rise velocity.

In Fig. 4.1, pictures of various bubbles sizes are shown and their rise velocity over their equivalent bubble diameter is depicted in the graph below. It is shown, that the curvature of the bubble interface is changing with increasing bubble volume and that concave and convex regions are formed during elongation. At $\lambda \approx 1$ the bubble shape (3) is totally convex, the minimal fluid film looks very thin and the rise velocity is the lowest ( $1^{st}$ local minima, $d_{eq} \approx 7.2$ mm, $v_B \approx 10$ mm s$^{-1}$). With increasing volume, the bubble tends to become more elongated, the shape is partly convex and partly concave, which leads to a thicker fluid film at the at all parts of the bubble, which results into a much higher rise velocity at $d_{eq} \approx 7.8$ mm ($1^{st}$ local maxima, $d_{eq} \approx 7.8$ mm, $v_B \approx 40$ mm s$^{-1}$). These two turning points are very important for the consideration of tubular reactors, because a slight difference in volume, can lead to a four times higher or lower rise velocity. A similar non-linear behavior of rise velocities is not known for any other bubbly flow, which makes Taylor bubbles a special candidate again, the core reason being the strong interaction with the capillary walls and the changes in the thickness of fluid film. With further increase of the bubble volume, the interface exhibits a second maxima, where the fluid film thickness becomes minimal again. This development is parallel to the second local minima in the rise velocity at $d_{eq} \approx 8.5$ mm, before the bubble reaches the Taylor bubble regime ($d_{eq} > 9.8$ mm, $v_B \approx 22$ mm s$^{-1}$). While increasing the bubble volume, there is no change in the fluid film thickness noticeable by eye anymore and the bubbles show a constant rise velocity as expected. The finding of this extraordinary slope of the rise velocity demanded a more profound investigation of the development in this regime. Therefore, data points close to the striking minima and maxima have been reproduced with two cameras, the first camera captures the bubble shape, while a second triggered camera with a higher magnification is focused on the bottom region of the bubbles, where the fluid film thickness $\delta_f$ is directly measured with a higher spatial resolution. In Fig. 4.1 the open diamonds indicate the reproduced bubbles and their terminal rise velocity, which are in good agreement with the original data (filled diamonds). The crosses mark the fluid film thickness $\delta_f$ on the secondary Y-axis, which is measured simultaneously. It is obvious, that the fluid film thickness and the rise velocity show the same trend as expected. This leads to the hypothesis, that the different regimes of small, intermediate and Taylor bubbles are explained by the minimal fluid film thickness and the shape of the bubbles. The reason why the bubbles in the $D = 7$ mm channel show this unique development, is that the intermediate and Taylor bubble interfaces do not show interfacial instationary waves. This is due to the high surface tension that enables a stable interface curvature and therefore, a steady fluid film thickness. The drag of the bubble strongly depends on the fluid film thickness, which leads to the remarkable and well defined minima and maxima in the terminal rise velocity.

In Fig. 4.2 the bubble rise velocity in dependency of the fluid film thickness $\delta_f$ of small, intermediate and Taylor bubbles is shown and the datapoints show a linear trend in between $260\mu$m$< \delta_f < 405\mu$m, where most values lie in between $\pm 20\%$ of the linear fit. This figure supports the hypothesis, that the bubble rise velocity is directly depending on the fluid film thickness. So there is an equilibrium between buoyancy force and drag force due to the bubble shape and therefore both determine the bubble rise velocity.
In Fig. 4.3, the measured values are compared to data from Clift et al. ( see also Fig. 2.10), where different rising bubbles in channels have been compared to find a general correlation for a normalized rise velocity in vertical channels in dependency of the diameter ratio of the bubbles [Cli78]. Therefore, the new data had to be adapted in two ways. Originally, the equivalent bubble diameter was used, which is not possible anymore for Taylor bubbles, because their diameter ratio would be larger than one. So the measured fluid film thickness at the bubble bottom was used to calculate the local bubble extension or 'radius' ($r_B = R - \delta_f$). Additionally, the original figure had to be extended, due to the relative small rise velocity ratio and the high diameter ratio of the new data, where the old figures has been cropped before. The data of bubbles rising in $D = 7$ mm vertical channel of this research (gray diamonds) have been added and fit very well to the existing correlation Eq. 2.35 (red line). Notice that this correlation was made by the investigation of small bubbles with a diameter smaller than the channel and which was now extend to intermediate and Taylor bubbles by using the local equivalent value of fluid film thickness to calculate the maximal local particle diameter $d_{max}$.

Due to the strong influence of the fluid film thickness on the global fluid dynamics of the bubble described by the terminal rise velocity, a strong influence on the local fluid dynamics of the fluid is assumed as well. But for such an investigation, tracer particles need to be added to the continuous fluid phase, to capture the effect of the fluid film thickness on the particle deposition in time.

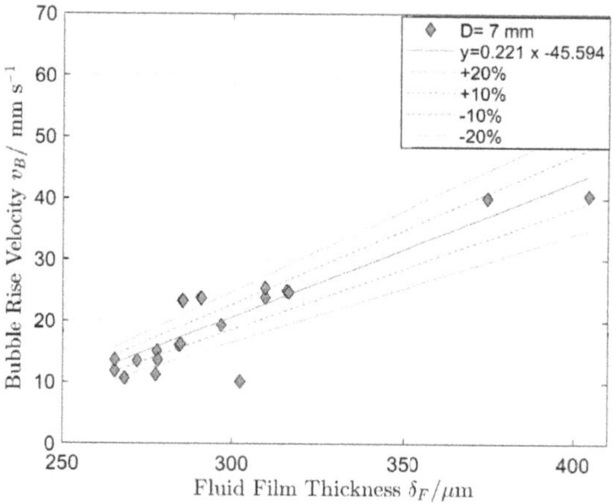


Fig. 4.2 Bubble rise velocity $v_B$ vs fluid film thickness $\delta_f$.

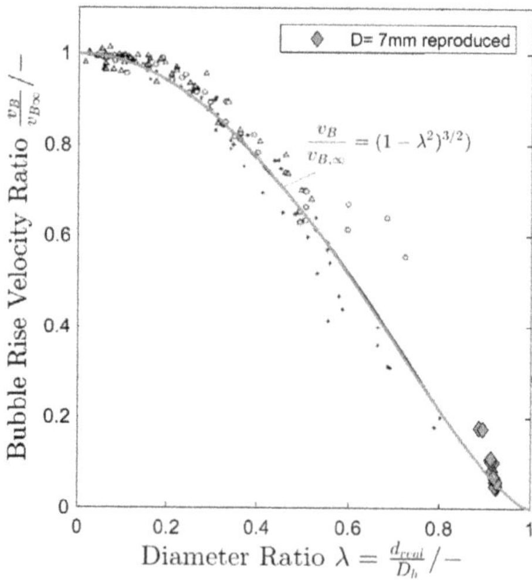


Fig. 4.3 Wall effects on bubbles and particles in channels. Rise velocity ratio $v_B/v_{free}$ vs. geometry ratio $d_{max}/D_h$. Extended figure to local maximal bubble diameter $d_{max}$, where the fluid film $\delta_f$ is the thinnest at the Taylor bubble bottom. Compare Fig. 2.10, after [Cli78] Note the extended axis.

Fig. 4.4 Average PIV images of small, intermediate and Taylor bubbles in a $D$ =7 mm channel in a counter current flow (average of 5 images at 200 Hz). The deposition of dispersed particles create pathlines similar to streamlines (if transitory stable flow structures are present), which enable the identification of characteristic wake flow structures like vortex cores and stagnation points.

**Local velocity field at small, intermediate and Taylor bubbles.**

The rise velocity of small, intermediate and Taylor bubbles rising in $D$ = 7mm channels shows strong minima and maxima in Fig. 4.1 caused by of the fluid film thickness. Especially, between the first minima and the first maxima the velocity becomes four times higher from about 10 mm $s^{-1}$ ($d_{eq}$ = 7.2 mm; $Re$ = 70) to nearly 40 mm $s^{-1}$ ($d_{eq}$ = 7.6 mm; $Re$ = 280). For a detailed analysis of the particle deposition bubbles with similar size were kept in the field of view by applying a counter current flow, which mean fluid velocity is similar to the rise velocity of the Taylor bubble. In Fig. 4.4 averaged raw PIV images (average of 5 images at 200 Hz) of bubbles from the different regimes are shown, where a) is a small bubble, b) & c) are intermediate bubbles and d) is a Taylor bubble. Due to the deposition of the tracer particles at long exposure times pathlines are distinguishable, which are similar to streamlines in case of stationary fluid flow. Therefore, region of higher velocities are indicated by longer traces and regions with very small particle displacement are indicated

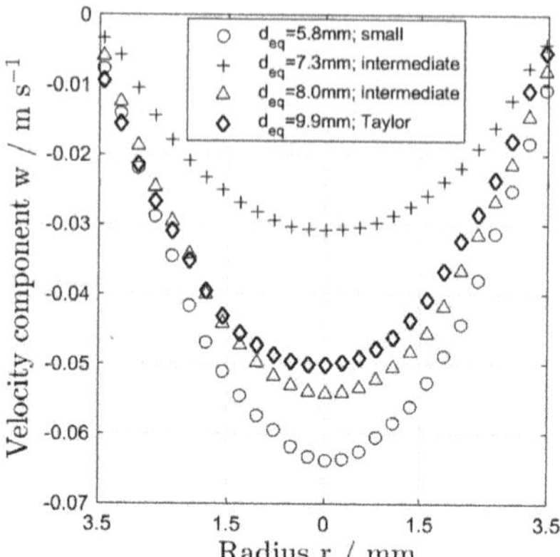


Fig. 4.5 Comparison of fluid inflow velocity profile taken with PIV measurements to keep small, intermediate and Taylor bubbles (from Fig. 4.4) fixed in a $D = 7$ mm channel by counter current flow.

by dots. From this figures, it is shown that the inflow above the bubble is laminar and a Hagen-Poiseulle-profile can be assumed (compare Fig. 4.4). Also, wake structures of the four bubbles show different sizes, shapes and symmetry, caused by the different rise velocities. For the case b) the intermediate bubble with the smallest rise velocity, the wake shows a triangle shaped region, where the local fluid velocity is small and this structure does not show vortices with local backward flows, here the flow field is similar to a jet. The three other cases show toroidal vortices with two vortex cores and stagnation points in the channel center, separating the flow within the wake in up-flowing and down-flowing fluid.

The small bubble in Fig. 4.4 a) ($d_{eq} = 5.8$ mm) should show a rise velocity about $v_B \approx 34$ mm s$^{-1}$ in stagnant liquid, which means that this velocity should be similar to the mean velocity of the fluid in counter current flow. In Fig. 4.5 the parabolic velocity profiles of the counter current flow above the bubbles are plotted, which are measured with PIV. It is known, that the mean velocity $w_{mean}$ of a laminar flow is half the maximum velocity in the channel center ($w_{mean} = w_{r=0;max}/2$). Therefore, the mean velocities of the counter current flow in Fig. 4.5 are similar to the rise velocity of the bubbles (Fig. 4.1), because the bubbles are fluid dynamically fixed in the observation field. Actually, the bubbles should rise a little faster than the mean velocity, because of the fluid film flowing down beside the bubbles. Therefore, the velocity of rising bubbles in stagnant fluid and of fluid dynamically fixed bubbles in counter current flow is not identical, but still gives a good impression of the fluid dynamics in this system.

It should be noted, that other fluid properties will change the characteristics of bubbles in channels. Therefore, it is not possible to establish a universal correlation with this data, but it becomes clear that the local minimal fluid film thickness defines the rise velocity of bubbles in vertical channels. This dependency is most significant for the investigated case $D = 7$ mm $Eo = 6.7$. In smaller channels $D < 7$ mm, there are smaller velocity differences between the minima and maxima and for larger channels $D > 7$ mm the bubbles showed already slight interfacial waves, which lead to unsteady bubble shapes and therefore unstationary fluid film thickness.
After the characterization of the flow regimes of small intermediate and Taylor bubbles inside the $D = 7$mm channel, the effect of channel size and geometry on the rise velocity will be analyzed in the following chapter.

### 4.1.2 Global bubble dynamics - Influence of the channel geometry

The shape and volume of free rising bubbles is affecting the rise velocity $v_B$ and the mass transfer across the interface as well. To characterize the flow regimes of bubbles in channels, it has to be distinguished between small, intermediate and Taylor bubbles, due to the slope of the rise velocity $v_B$ vs. equivalent bubble diameter $d_{eq}$, which is actually caused by the fluid film thickness $\delta_f$. It is important to identify, which equivalent bubble diameter is the critical bubble diameter to determine the Taylor bubbles mass transfer. This is important especially for the following global and local mass transfer investigations, where the fluid dynamic condition should not change during the bubble dissolution to obtain reproducible measurements. Therefore, Taylor bubbles with their volume independent rise velocity are the ideal tool for detailed mass transfer analysis. In this Chapter the rise velocity of bubbles rising in different channel diameter and cross sections will be investigated.

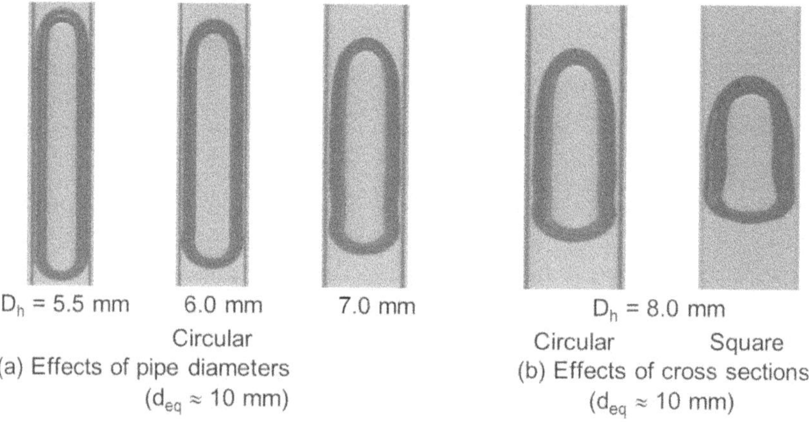


Fig. 4.6 Bubble shapes for a constant bubble volume in various channels. (a) Effects of pipe diameter ($d_{eq} \approx 10$ mm; $5.5 < D < 8$ mm), (b) effects of channel cross section, circular and square ($d_{eq} \approx 10$ mm; $D_h = 8$ mm).

Fig. 4.6 a) shows a comparison of bubble shapes each $d_{eq} \approx 10$ mm) in channels in function of several hydraulic diameters $D_h$ and channel shape (Fig. 4.6(b)). Obviously, the liquid film thickness increases with $D_h$, where a higher rise velocity is expected. For each channel several bubbles have been measured to be able to find the characteristic Taylor bubble regimes. The bubble length and liquid film thickness of the bubble in the circular channel are different from those in the square channels even for the same $d_{eq}$ and $D_h$.

**Terminal rise velocity - the effect of the hydraulic channel diameter**

Figure 4.7 shows a comparison of the bubble rise velocity $v_B$ for various air bubble volumes in channels with different hydraulic diameter $D_h = 5.5$ mm, 6 mm, 7 mm & 8 mm (filled circles, open triangles, filled diamonds & open squares). The terminal rise velocity of Taylor bubbles is independent of $d_{eq}$ and it decreases with decreasing $D_h$. This is because the buoyancy force is gradually overcome by the interfacial tension force as $Eo_D$ decreases with $D_h$. The rise of elongated bubbles in vertical pipes would cease at $Eo_{Dcrit.} = 4$, at which $D_{crit.} = 5.43$ mm in a aqueous system. Since the Taylor bubbles in the $D_h = 5.5$ mm pipe are very close to $Eo_{Dcrit.}$, their constant rise velocities are as small as $v_B \approx 2$ mm s$^{-1}$ and the fluid film is barely visible.

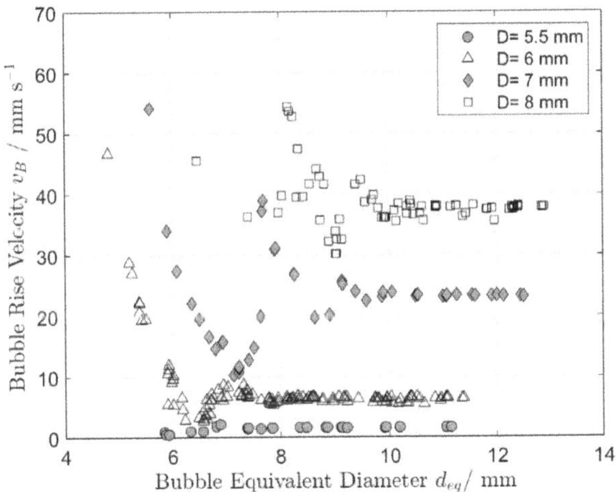


Fig. 4.7 Bubble rise velocity $v_B$ vs. bubble equivalent diameter $d_{eq.}$ in channels with different diameter $D$. Flow regimes of small, intermediate and Taylor bubbles can be identified in all channels $5.5 < D < 8$ mm (data from [Kas15])

Additionally, in Fig. 4.7 the increase of the terminal rise velocity of Taylor bubbles with increasing channel diameter is shown. Taylor bubbles in $D = 8$ mm are rising nearly 20 times faster than in $D = 5.5$ mm. The strong increase can be explained with the ratio of buoyancy forces to interfacial forces of the fluid system, which usually is described with the Eötvös number $Eo = g\,\Delta\rho\,D^2/\,\sigma$. For bubbles in $D = 5.5$ mm and 6 mm the surface tension is dominant due to the small curvature radius of the interface. The surface tension tends to generate spherical bubbles and press against the fluid film, while the buoyancy force tends to elongate the bubbles. For bubbles in the $D = 8$ mm channel, the influence of the surface tension is reduced, because the curvature radius is larger, according to Laplace. This dependency of the Taylor bubble rise velocity on the channel diameter can be used to adjust and control the fluid dynamic condition of an experiment by choosing the ideal channel diameter. This is an high advantage compared to free rising bubbles, where usually the physical properties need to be adapted, which effects the coupled transport processes. Additionally, it is observed that the interfaces of the bubbles are not steady anymore and slight interfacial waves can occur on some bubbles. That is why the intermediate regime is not so well defined as for the bubbles in the $D = 7$ mm channel.

#### Terminal rise velocity - the effect of the channel cross section (square and circular)

The bubble rise velocities for various bubble volumes were measured to obtain a range of constant rise velocity of Taylor bubbles in square and circular cross section. Fig. 4.8 shows a comparison of $v_B$ for various bubble volumes in square (marked as squares) and circular channels (marked as circles) with the hydraulic diameter $D_h = 6$ mm (filled dark) & 8 mm (filled bright). In the $D_h = 6$ mm channels the three regimes of small, intermediate and Taylor bubbles are significant, even if the overshoots are less significant. In the $D_h = 8$ mm channels, the regime of small bubbles is not shown, because of dominant dynamic shape oscillation, which lead to high errors in the estimation of $v_B$ and $d_{eq}$ for an analysis with only one camera perspective. The intermediate and the Taylor bubble regime can both be determined. Due to the non-rotational symmetric geometry in the channel with square cross section, the bubbles and the fluid film are not symmetric either. Therefore, a detailed analysis of the fluid film thickness is not suitable with back-light imaging techniques and only the rise velocity and the bubble equivalent diameter are analyzed. The focus of this chapter is on experimental data in which the rise velocity $v_B$ of Taylor bubbles are constant. The rise velocity is independent of $d_{eq}$ and it decreases with decreasing $D_h$. Due to the same reason explained before for circular channels. The rise velocities of bubbles in the square channels of $D_h = 8$ and 6 mm are more than 1.5 and 3 times higher than those in the circular channels of $D_h = 8$ and 6 mm. This can be explained by the presence of large flow areas for the liquid phase in the corners of the square cross section, while the bubble cross section is in an intermediate state between circular and square. A larger cross sectional area between the bubble and the channel wall is providing a higher volume flow rate within the film around the bubble. With the same fluid dynamic boundary condition at the wall and at the fluid interface, this leads to higher rise velocities of the bubbles in all regimes. These fluid dynamic flow conditions might affect the global and local mass transfer from $CO_2$ Taylor bubbles, due to high local velocities and enhanced convection.

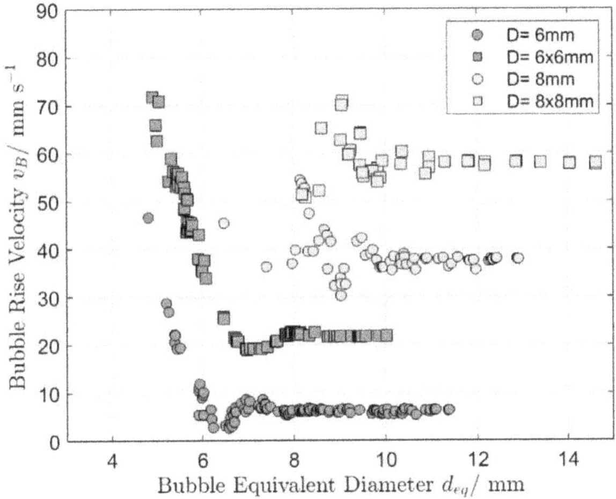


Fig. 4.8 Comparison of bubble rise velocity $v_B$ vs. bubble equivalent diameter $d_{eq.}$ in square and circular channel cross section $D_h = 6$ and 8 mm.

It is noted, that bubbles in square channels are more unstable and show more intensive shape deformation and interfacial waves than bubbles in circular channels. That is the reason why they are more difficult to be kept in a counter current flow, where the fluid film can rapture more easily at one point due to a slight difference in the velocity and pressure field. But in general it is possible to determine the Taylor bubble regime so investigations on mass transfer processes and bubble dissolution are possible without major variations of the bubble rise velocity. In the following fluid motion around air bubbles in counter current flow needs to be characterized and discussed, before the gas phase is changed to $CO_2$ and mass transfer processes are taken into account.

### 4.1.3 Local fluid dynamics at bubbles in channels - Influence of the channel geometry and fluid properties

For a deeper understanding of the coupling of transport processes at fluid interfaces, the liquid flow is analyzed via PIV methods. In the following, the velocity field around bubbles in vertical channels will be analyzed in detail, shortly introduced by averaged raw PIV images in Fig. 4.4. Also the effect of channel geometry, with the variation of diameter and cross section, on the general flow field will be presented for small and Taylor bubbles. It is noted, that the experiments were executed in counter current flow and stagnant liquid depending on the purpose and stability of the fluid film.

**Influence of the hydraulic channel diameter $D_h$ on local flow field and wake structure.**

Taylor bubbles are kept in counter current flow, where the velocity profile (component w in vertical direction z) far upstream of the bubble is typically a parabolic laminar Hagen-Poiseuille flow, due to the low rise velocity and Reynolds number of bubbles in channels as introduced in Fig.4.9. The maximal velocity $w_{max}$ of each velocity profile occurs in the center $w_{center,r=0}$ in Fig. 4.9 and it is increasing with increasing channel diameter $D = 6, 7 \& 8$ mm as expected from the Taylor bubble rise velocity Fig. 2.12. The general flow characteristics around a bubble in a channel will be determined for small air bubbles in counter current

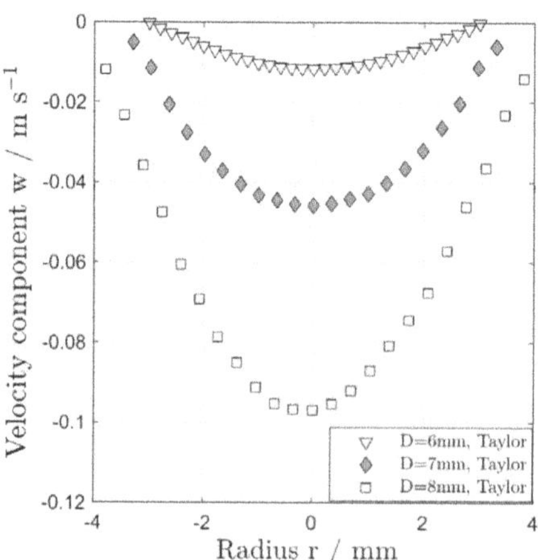


Fig. 4.9 Comparison of fluid inflow velocity profiles $w(r)$ measured by means of PIV to keep Taylor bubbles fixed in channels $D = 6, 7 \& 8$ mm by counter current flow ($Re$ = 36, 153, 303).

flow, because here the compromise of field of view size and spatial resolution is optimal for PIV investigations. The field of view captures a small bubble itself and enough space in front and behind to identify the velocity field regions which are influenced by the bubble. In Fig. 4.10 (left) and 4.11 (left) the velocity fields around small air bubbles fixed in a counter current flow in a $D = 6$ mm & $D = 7$ mm circular channel is shown. The bubbles are in the center, where the bubble tip is indicated by the local z coordinate $z^* = 0$ mm. The parabolic inflow can be seen far in front of the bubble and after each bubble a wake structure is formed by the disturbance of the film flow. With decreasing distance from the bubble front, the velocity profile is deformed by the bubble even upstream of the bubble ($z^* > 0$). This can be seen easily in Fig. 4.10 (right) & Fig. 4.11 (right), where six or seven velocity profiles along z-direction are plotted at different constant radiuses $r$. In both figures, the thick blue profile indicates the center velocity $v(r = 0$ mm) and deviation from Hagen-Poseuille can be distinguished at a height $z^* = 5 - 6$ mm, due to the fluid separation into the fluid film between bubble and wall. At $z^* = 0$ mm the cross sectional area of the fluid decreases, due to the bubble presence, which leads to an acceleration of the fluid, due to the continuity of mass, to even higher local velocities than $v_{max}$ inflow (compare $D = 6$ mm; $r = 2.5$ mm; 0 mm $< z^* < -7.5$ mm). With beginning of the deceleration of the fluid close to the channel center, the fluid closer to the wall is being accelerated due to flow redirection. Additionally, both bubbles have a stagnation point (Fig. 4.10 (left) and 4.11 (left) dark blue) at $z^* = 0$ mm and $r = 0$ mm, where the vertical flow is totally converted into an horizontal flow direction to pass the bubble.

Behind both bubbles the wake structure begins, which is defined by the bubble induced flow disturbance until the laminar parabolic inflow recovers. When comparing different channel sizes, it is obvious, that the wakes are very different in the length, structure and phenomenology. While in the $D = 6$ mm channel, the wake structure looks laminar, with a steady acceleration in the center $r = 0$ mm, the profile in $D = 7$ mm channel shows a local backward facing flow at about $r = 0$ mm; $z^* = -10$ mm, much like assumed toroidal vortex. This vortex can be identified by the two vortex cores at about $r = \pm 2$ mm; $z^* = -8$ mm and the stagnation point or hyperbolic point at about $r = 0$ mm; $z^* = -12.5$ mm, where the horizontal flow from the outside vortex surface is redirected into vertical flows into the inner vortex (up-flow) or into the channel outflow (down-flow). This three significant points are all determined by the low local velocity amplitude and the characteristic velocity directions (rotation around two vortex cores ; hyperbolic flow at stagnation point). The length of wakes can be estimated, where the the vertical velocity $w$ recovers 95% of its initial value $w(r = 0\text{mm}) > 0.95 w_{max}$. In the $D = 6$ mm channel (Fig. 4.10) the flow is nearly recovered after 1.5 $D$ behind the bubble.

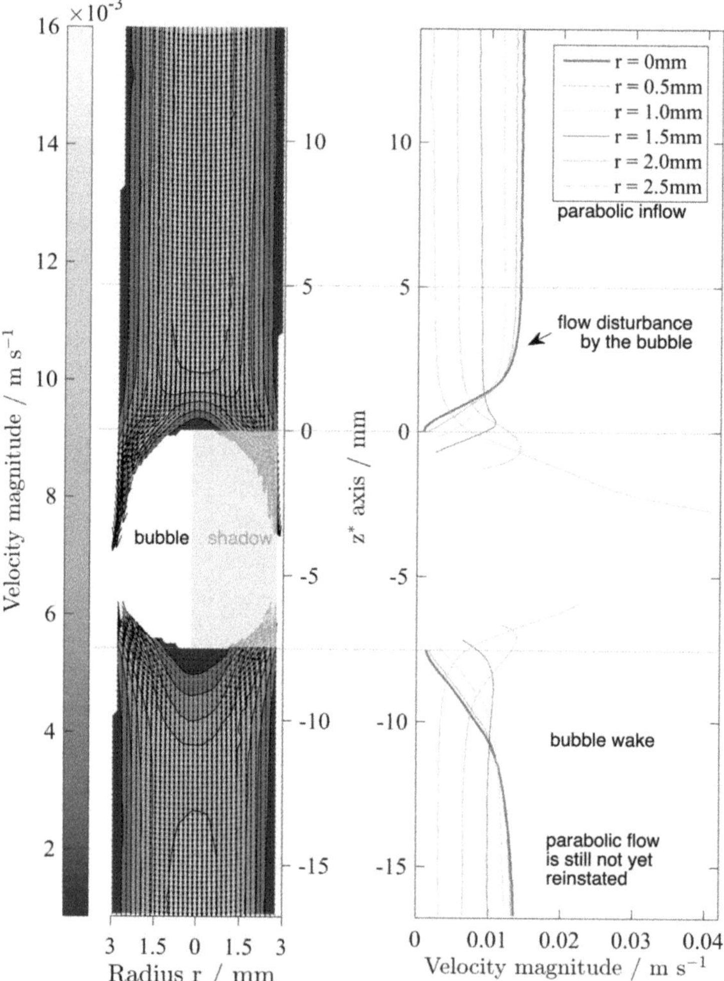


Fig. 4.10 Local fluid velocity around a bubble in the $D = 6$ mm channel in counter current flow measured by means of PIV. Influence of the bubble on counter current inflow, the wake structure and the flow recovery.

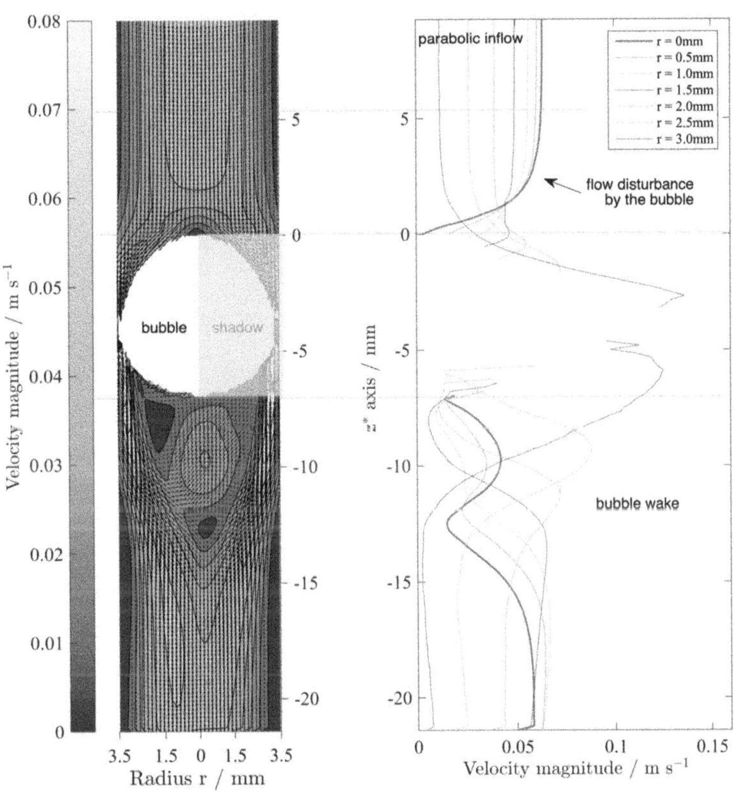


Fig. 4.11 Local fluid velocity around a bubble in the $D = 7$ mm channel in counter current flow measured by means of PIV. Influence of the bubble on counter current inflow, the wake structure and the flow recovery.

The wake in the $D = 7$ mm channel is still far away from the parabolic inflow even after $2D$ behind the bubble. Additionally to the velocity profile along the z-direction, the profiles along the radius in the $D = 6$ mm and $D = 7$ mm channel are plotted in Fig. 4.12 and 4.13. On the left, the velocity profile development from parabolic inflow to the accelerated flow inside the fluid film is shown. On the right, the development from the accelerated fluid film flow, the wake flow and to partly reinstated parabolic flow is shown. The profiles are taken equidistant from the velocity field presented in Fig.4.10 (left) and 4.11 (left). The non symmetric peaks in the fluid film are mainly caused by errors in the PIV post processing induced by the bubble shadow inside the laser sheet. Therefore, it can be assumed that the left part of the flow field is more reliable. In general, much higher velocities are measured together with high local

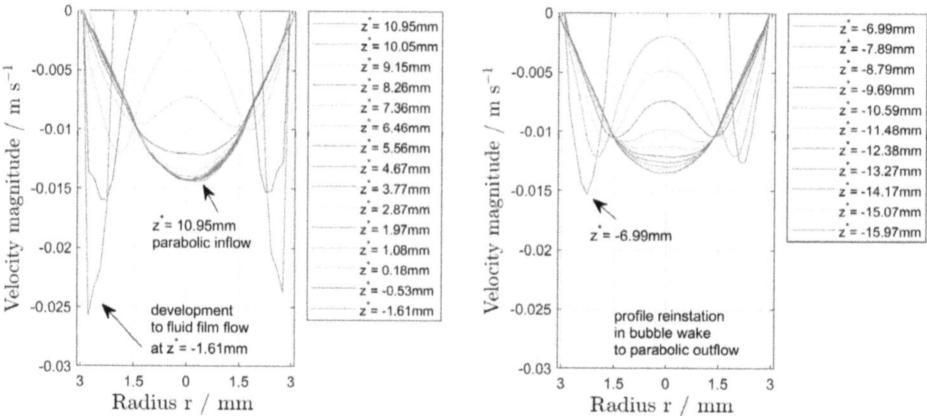


Fig. 4.12 Local fluid velocity profiles along radius for constant z-position at a bubble in $D = 6$ mm channel in counter current flow. Left: Evolution from parabolic inflow to accelerated film flow. Right: Evolution from accelerated film flow to the wake and recovered parabolic flow.

velocity gradients in the film between bubble and wall, compared to the laminar counter current flow. This already shows the dilemma of multiscale processes in between the demand of an analysis of the overall Field of View and the high spatial resolution for the PIV analysis of local phenomena. Also the high differences in the velocity magnitudes require different spatial and temporal resolution to resolve the fluid dynamic problem in an ideal way.

In the following, the local flow field at the bubble front is in focus, with another experimental setting to enable a higher spatial resolution. As explained before, a higher spatial resolution is needed, due to the evolution of the velocity field along the interface of the bubble front, where the velocity gradient increases and leads to a higher shear rate at the

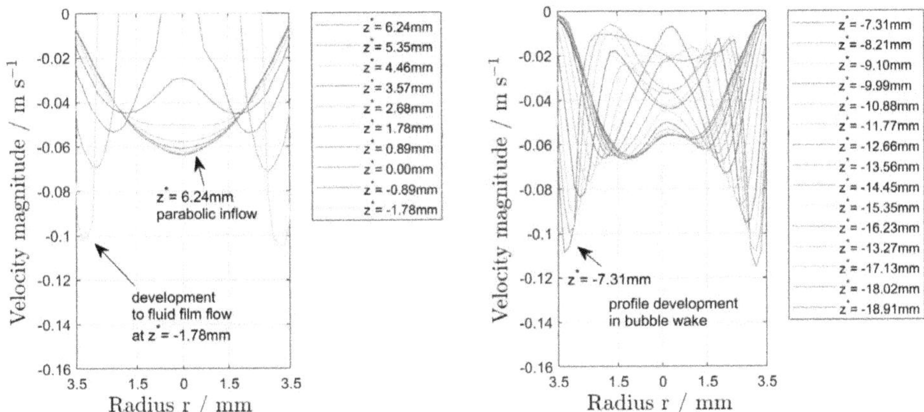


Fig. 4.13 Local fluid velocity profiles along radius for constant z-position at a bubble in $D = 7$ mm channel in counter current flow. Left: Evolution from parabolic inflow to accelerated film flow. Right: Evolution from accelerated film flow to the wake.

interface. Even with this higher spatial resolution in Fig. 4.14 or Fig. E.1 it is noticeable, that the spatial resolution with PIV is too low to measure the expected no-slip boundary condition at the wall $w(r{=}3\text{ mm}){=}\ 0$ m s$^{-1}$. The velocity gradient of the parabolic inflow of 0.01 m s$^{-1}$ / 3 mm at $z^* = 2.5$ mm changes dramatically by a factor of about $f = 15$ when turning into the film flow of 0.01m s$^{-1}$ / $< 200\ \mu$m at $z^* = -0.5$ mm. The velocity magnitude and therefore the velocity gradient gets even stronger far in the film flow, but this is not measurable anymore with this setting (compare Fig. E.2). Actually, the magnification and the particle seeding needs to be increased with even smaller particles to be able to choose even smaller grids for the post processing with cross correlation, otherwise high gradients within this thin fluid layer $\delta_{film} = 300\ \mu$m will always lead to averaged velocity values, which are partly overestimated for low and underestimated for high actual values. Although it is obvious that the tangential velocity along the interface is changing dramatically from the tip to film region. Here, it can be assumed, that this could effect the local convective mass transfer at the interface and the concentration boundary layer thickness $\delta_c$ along the interface.

In Fig. 4.14 the flow structures at the bubble front between the channel center and the wall are shown again, to focus on the evolution along the interface. This flow structures with a relative large zone of low fluid velocity at the stagnation point is unusual for free rising bubbles, where the high shear flow of the fluid passing the bubbles, induces a high internal gas circulation inside bubbles. The missing wall influence of free rising bubbles let the fluid flow easily around the bubble. The phenomena of relatively large zones of low fluid velocity,

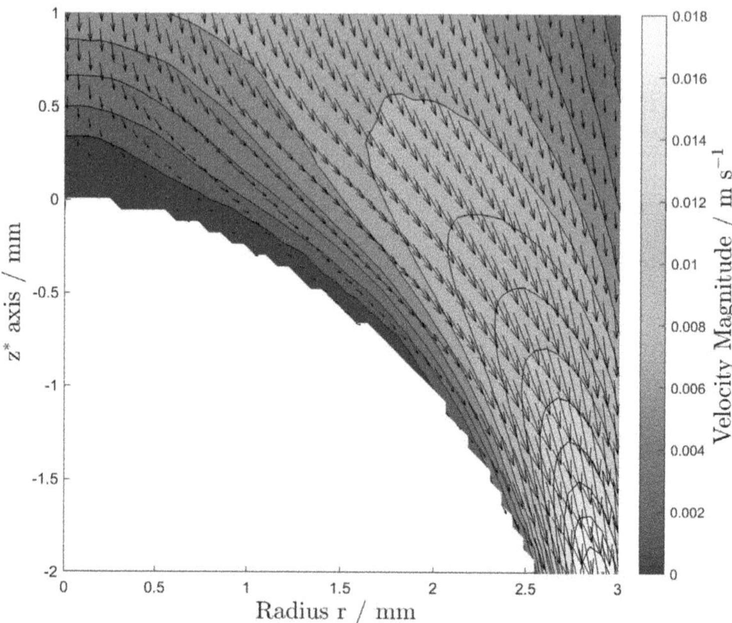


Fig. 4.14 Local velocity field at a Taylor bubble front kept in counter current flow within a $D = 6$ mm circular channel. Fluid flow acceleration due to the reduction of available cross sectional area caused by the bubble interface.

is special for Taylor bubbles in small channel diameters. The tangential fluid flow, would transport a transferred species away from the local interface into the bulk phase. Due to the strong increase of the fluid velocity in around the Taylor bubble it is expected, that the fluid dynamic boundary layer decreases from the bubble front into film region, which also should lower the mass transfer resistance at his point. However, interestingly, the fluid directly at the interface is not the fastest, which could be explained with an immobilization of the interface by surface active agents which prevent a slip condition but more reasonable is, that internal gas circulation, which is induced by the friction of the liquid phase, establish two or more counter rotating vortices; at least an elongated toroidal vortex in the center of the bubble and a smaller toroidal vortex in the Taylor bubble tip. But without numerical simulations this assumption cannot be clarified. Any how, the velocity field at the interface of the Bubble front shows the increase of the velocity gradient along the interface into the fluid film. On one hand, this should enhance the local mass transfer performance $\mathrm{Sh}_{loc.}$ according to the Two Film Theory ant the analogy of the transport processes along the interface. On the other hand, the fluid elements, assumed by the Surface Renewal Theory, moving along the interface have a larger contact time $\tau$ due to the slow flow field a the front. This could be compensated by the increasing circumference of the bubble, which would lead to a stretching of the existing fluid elements or a refill by fresh fluid elements from the bulk (convectional transfer).

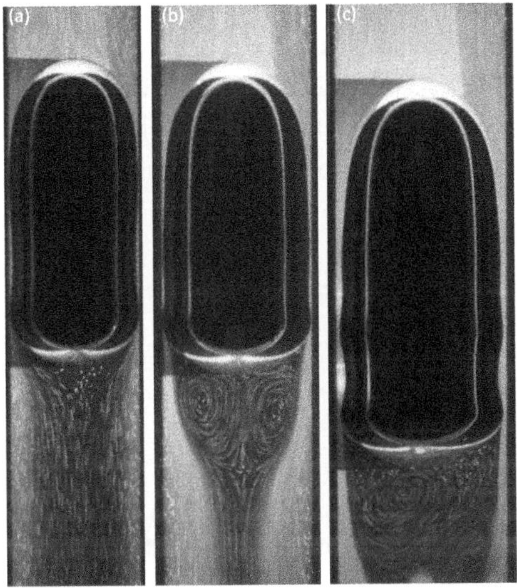


Fig. 4.15 Averaged raw LIF/PIV images generating particle pathlines and the regions of bulk (bright) and wake (dark areas) for Taylor bubbles in counter current flow in the a) $D = 6$ mm, b) $D = 7$ mm, c) $D = 8$ mm channel [Kas17b].

Behind the Taylor bubble the fluid is coming from the liquid film region and decelerates again due to the increasing cross sectional area for the fluid. Far away form the bubble, its effect on the fluid flow will be neglectable and the parabolic flow will reinstate again. The position of the flow recovery sets the endpoint of the wake region which usually depends on *Re* and *D* as shown in Fig. 4.9 and Fig. 4.15. In the averaged raw PIV images (averaged 5 images at 200 Hz), pathlines of the tracer particles indicate fluid streamlines (static flow assumed) and therefore already qualitatively the wake structure behind the different Taylor bubbles in counter current flow in different channel diameters $D$= 6, 7, 8 mm. It is obvious that with increasing channel diameter, (and therefore, Eötvös number, rise velocity and Reynolds number), the wake is changing from a laminar creeping flow, to a toroidal vortex, to an open turbulent mixed wake with dynamic vortex shedding and three dimensional movement. The wake structure in the $D$ = 6 mm channel in Fig. 4.16 shows barely any backmixing with a small area of a deadzone in the center of the channel close to the bubble bottom. The closed structure of the toroidal vortex in the $D = 7$ mm channel shows a coherent structure with a very stable vortex and a very strong backmixing, which is longer than the channel diameter. In the $D = 8$ mm channel the mixing within in the wake seems to be even stronger, due to the velocity fluctuation of turbulent mixing and vortex shedding. The position of the core vortex is fluctuating more strongly and is less steady. The shedding of the vortices can be seen by the strong asynchronous velocity field between the sides. Upper left shows a strong vortex core and in the lower right wake a weaker small vortex is just divided from the rest of the wake and transported away in the bulk in Fig. 4.15 & Fig. 4.16 .

The dependency rise velocity and wake structure behind Taylor bubbles from the channel diameter could help to understand and to adjust mixing timescales with the bulk for reactive Taylor bubbles.

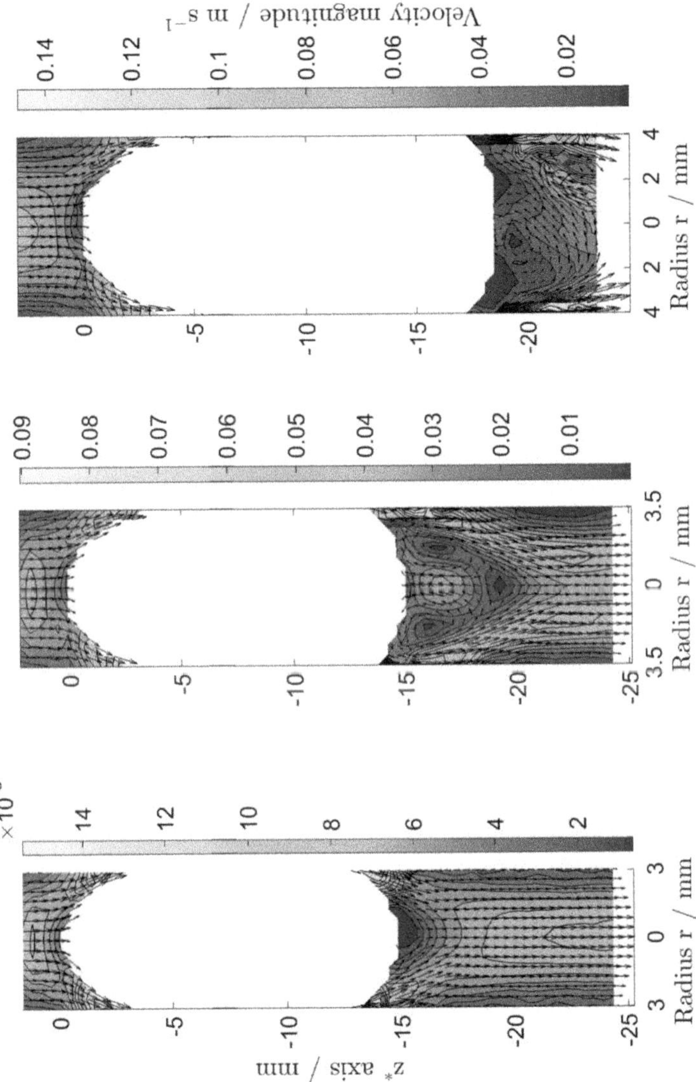


Fig. 4.16 Velocity fields around Taylor bubbles kept in counter current flow in three different channel a) $D$ = 6 mm, b) $D$ = 7 mm, c) $D$ = 8 mm ($Re$ = 36, 153, 303 ). Note the color-map legend with significant different fluid velocity ranges.

**Local flow field influenced by the fluid viscosity in square channels**

The geometrical differences in the channel cross section induce large differences in the local flow field, due to the wall effects and shear flows in single phase but also in multiphase systems like Taylor bubbles. In channels with a circular cross section, the wall effects are rotation symmetric, which results in the well known parabolic Hagen-Poiseuille velocity profile for single phase flow. This rotation symmetry causes also a symmetric pressure field, which stabilizes the Taylor bubble shape. Therefore, it is obvious, that the asymmetry of square cross sections affects the single phase flow field, but especially the multiphase system.

The measurement procedure needs to be applied to the square cross section by measuring in several parallel planes successively to get deeper insights into the 3D flow structure. To enable this measurement, the channel was moved out of the center with a precise linear table, as shown in Fig. 3.9. Especially in the wake region the 3D flow structure is dominant in aqueous systems with low liquid viscosity. Rotation symmetrical flows in circular channels lead to stabilized wake structures, with restrained mixing compared to the square channel. In square channels local bypass flows in the corners occur, which lead to pressure gradients around the bubble circumference, which influence the bubble shape, depending on the fluids interfacial tension or $Eo$. Additionally, it was shown in Fig. 4.8, that Taylor bubbles in square channels rise much faster than Taylor bubbles in circular channels. Therefore, $Re$ increases, which enhances the turbulence in the wake, where longer and open wake structures with higher rates of vortex shedding are expected. Additionally, the bubbles in counter current flow in square channels are more likely to touch the wall, by rupturing fluid film at the channel wall. Therefore, bubbles in square channels within aqueous solutions are less stable in shape and long-term observations are more difficult to perform. To stabilize the bubbles and to increase the fluid film thickness, aqueous solutions with glycerol are prepared with enhanced fluid viscosity.

In Fig. 4.17 the velocity fields around one particular small air bubble in four parallel planes at $y = 0$ mm (centered), $y = -1, -2$ and $-2.8$ mm (all acentric) are shown. Note that the planes cut the bubble less with increasing distance within $y$. Due to the asymmetry of the channel, the four velocity fields look very different. The centered 2D flow field at $y = 0$ mm shows a quite low velocity magnitude of about $w = 0.04$ m s$^{-1}$, but with increasing $y$ the bypass flow in the channel edges becomes more dominant, where the fluid flow velocity magnitude becomes more than three times higher. The comparison of the mean inflow velocity shows an evolution in the opposite direction, where at $y = 2.8$ mm the plane is very close to the wall, where the no-slip condition and the viscosity of the fluid reduces the flow velocity in this plane, the inflow velocity in the centered plane shows nearly parabolic flow structures. The wake region shows a slight asymmetric flow structure, which can be caused by an insufficient

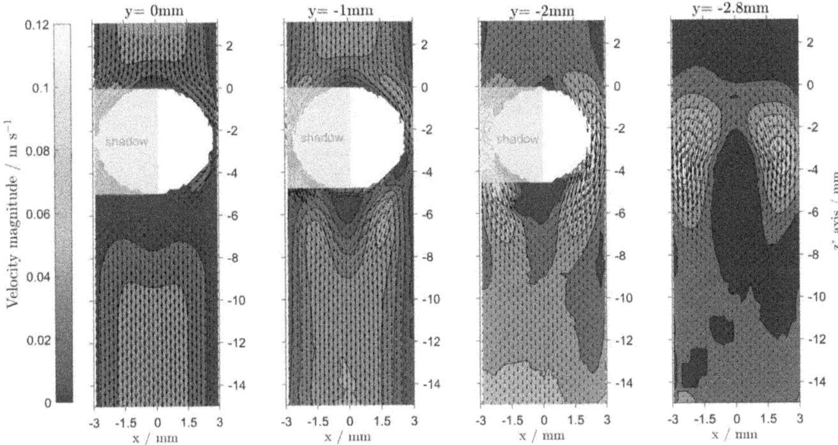


Fig. 4.17 Measured mean 2D velocity fields in four parallel planes at one particular small air bubble kept in counter current flow in $D_h$ = 6 mm (glycerol 58 % water 42 % solution )

time period of averaging of the PIV images. Due to the strong bypass flow in the edges of the channel, there is another phenomena, which deserves more attention. Fig. 4.18 a) and b) show the flow field and the flow profile of the plane at $y = 0$ mm. [3] The high shear gradients around the bubble circumference, lead to a complex internal gas re-circulation flow, which induces fluid motion at the wall faces of the channel. But the direction of the fluid flow is upwards and therefore against the main flow direction, which can be seen easily by the change of the local fluid velocity component $w$ from negative to positive sign close to the bubble interface ($-2.5 < z^* < -5$ mm; $y = 0$ mm). Additionally, the magnitude of the upwards flow is of comparable strength with $w(r = 3\text{ mm}) > 0.02\text{ m s}^{-1}$ which can be seen in Fig. 4.18 b), where velocity profiles along the z-axis for constant $r$ are shown. The spatial evolution of the velocity profiles of the inflow and the wake structure is shown by the deceleration of the velocity profiles along different constant radius $r$ in-front the bubble and the acceleration and slow recovery of the flow profile in the wake structure.

In Fig. 4.19 the velocity profile along the main axis of the square cross sectional channel shows a typical parabolic inflow far away from the Taylor bubble front, like in single phase laminar flows. The influence of the bubble on the velocity field is not pronounced in-front of the Taylor bubble. Additionally, the fluid in the central planes of the channel is accelerated towards the corners of the channel, due to the thicker fluid film and larger area, where less resistance is present. Therefore, the acceleration out of the center plane in the main axis is

[3] The same comparison of the other planes $y = -1$, $-2$ and $-2.8$ mm can be seen in the appendix F.1-C.9.

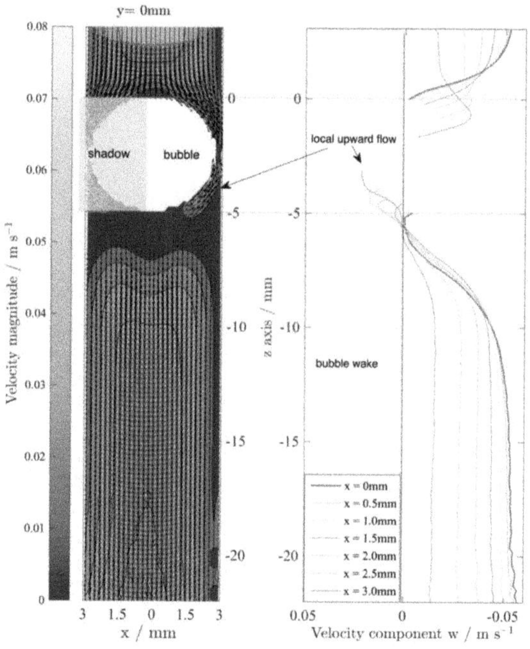


Fig. 4.18 Measured mean 2D velocity fields at $y = 0$ mm at a small air bubble kept in counter current flow in the $D_h = 6$ mm square channel.

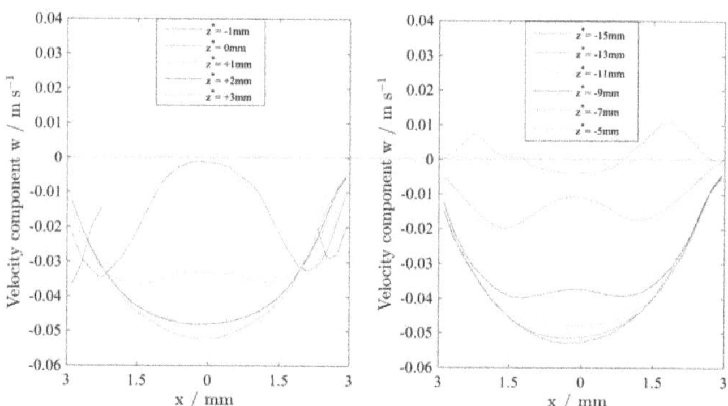


Fig. 4.19 Velocity profiles at $y = 0$ mm and different heights $z$ in-front (left) or behind (right) a small air bubble kept in counter current flow in the $D_h = 6$ mm square channel.

much stronger than in the circular channel. This leads to very different velocity gradients at the bubble interface along the rotation angle around the channel center. Similar flow phenomena have been reported by Aland et al. and Meyer et al. in pressure driven Taylor flow bubble chains in $D_{hyd} = 2$ mm using PIV measurements and CFD [Ala13] [Mey14].

**Wake length of rising Taylor bubbles and the influence of fluid viscosity**

PIV measurements in the channel center ($y = 0$ mm) of the $D_h = 6$ mm square channel have been made to estimate the effect of viscosity over the wake length behind the rising Taylor bubbles. Therefore, Taylor bubbles in aqueous solutions of viscosities $\nu = 1, 2, 4$ and 8 mPa s have been investigated, to show the qualitative wake structure and to quantify its length within the liquid phase, which gives an idea about the qualitative energy dissipation from motion to heat. The estimation of the wake length is shown exemplary in Fig. 4.20, where the end of the wake is defined, by the distance between the bubble bottom $z^* = 0$ mm and the position, where the velocity magnitude of the fluid at the channel center falls below 5% of the rise velocity of the Taylor bubble $v_B(\nu = 8 \text{ mPa s}) = 22 \text{ m s}^{-1}$.[4] Even though, the fluid density $\rho$, viscosity $\nu$ and surface tension $\sigma$ of the glycerol-water solutions are depending on the solutions glycerol proportion, the rise velocity of the Taylor bubbles are similar within 10 % deviation. The rise velocity, fluid properties, dimensionless groups and the estimated wake length are shown in Fig. 4.21. As all Taylor bubbles rise with a similar velocity, the induced energy into the fluid phase should be similar as well. The influence of the glycerol concentration on the bubbles Reynolds number and estimated wake length show similar behavior in Fig. 4.21, bottom. With increasing glycerol concentration in the solution, the density and the viscosity of the solution is increasing, which affects the main dimension less numbers. Therefore, inertia and the viscous forces are increasing, which influences the energy dissipation rate in the fluid. Therefore, the wake length is strongly decreasing with increasing viscosity. The term of dissipation of Navier-Stokes equation equals the first derivative of the shear stress $\tau_{zx}$ in main flow direction it follows

$$\tau_{zx} = \frac{F_F}{A} = \eta \cdot \frac{dv_z(x)}{dz} \tag{4.1}$$

The friction force $F_F$ in between the fluid layers is directly proportional to the dynamic viscosity $\eta$, where the energy dissipates into heat. Therefore, it is obvious that the wake length and *Re* are decreasing with increasing viscosity of the fluid, like seen in Fig. 4.21. This relation between the viscosity and the energy dissipation is not surprising or new, but the

[4]The wake lengths behind the Taylor bubbles rising in the other aqueous solutions are shown in the appendix G.1-E.5.

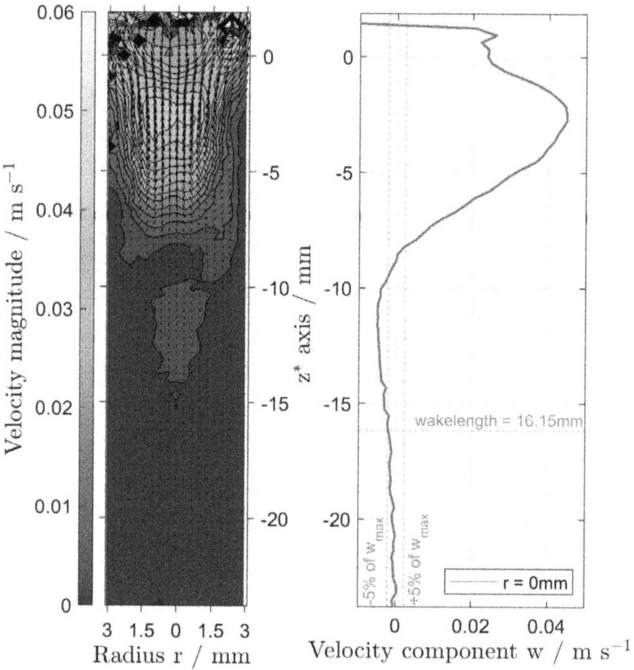


Fig. 4.20 Estimation of the wake length behind a rising air Taylor bubble in a square channel $D_h$ = 6 mm in the centered 2D plane. Bubble bottom at $z^* = 0$ mm.

possibility to measure and proof this inside a two phase flow system is an unexpected fortune. Usually, changing fluid parameters induce a change in bubble shape and therefore rise regime for free rising bubbles, where qualitative differences are so strong, that local quantitative comparison of the flow field is useless. Taylor bubbles overcome this problem for this two phase system and even quantitative comparisons become feasible, due to the qualitative similarity of shape and rise velocity of the Taylor bubble. In this case, the wake region behind the bubble, where the backmixing is taking place is reduced strongly, compared to the pure water. This could be an interesting research approach for reactive processes, where the rise velocity stays constant but due the control of the viscosity the mixing timescales can be adapted to the reactive time scales.

Even though, the flow structures in square channels show interesting phenomena, the analysis of the velocity field in different parallel planes at a particular bubble shows that the flow structure is so complex that reproducible experiments with a high temporal and spatial

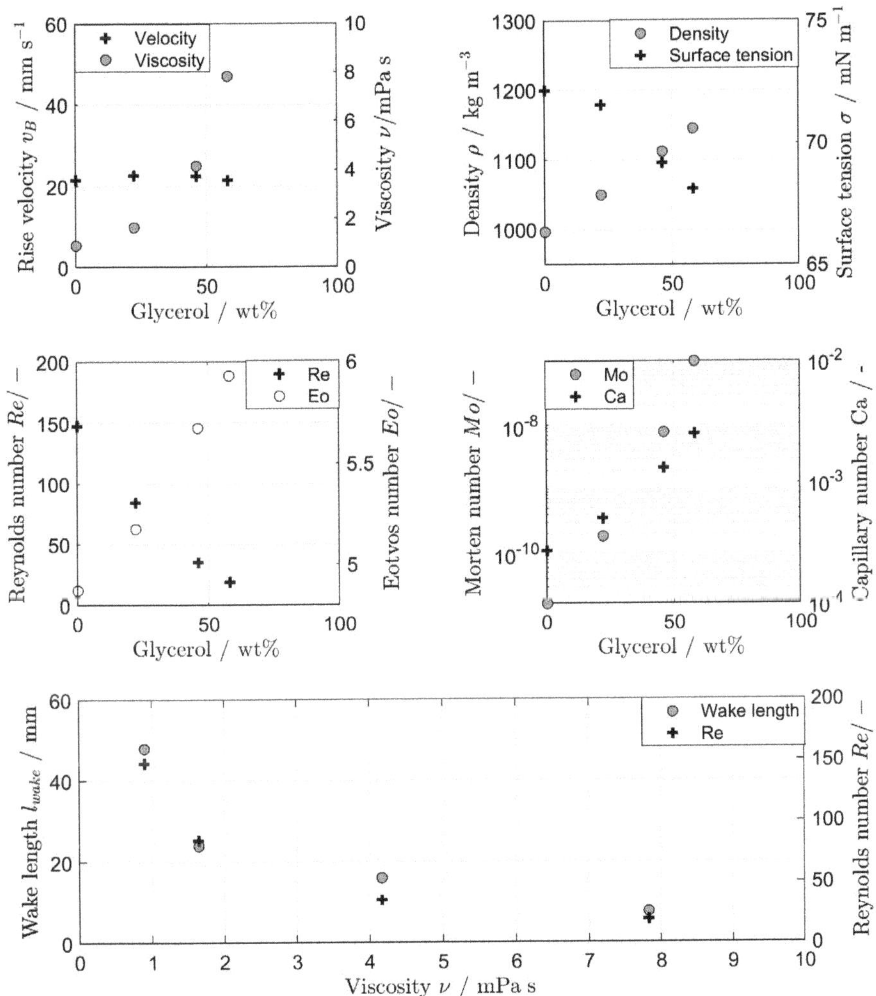


Fig. 4.21 Wake lengths behind air Taylor bubbles in the square channel of $D_h$ = 6 mm, rise velocity, fluid properties and dimensionless numbers for different viscous aqueous solutions.

resolution are unfeasible to identify direct links between the local and global fluid dynamics and mass transfer. It is not possible to directly measure local parameters in parallel planes simultaneously without a high-speed scanning technique, which would be necessary to

understand the coupling of mass transport across the interface the convection and mixing into the bulk liquid phase. Especially, the comparison of the velocity field in all four planes in the appendix Fig. F.1-C.9 shows in an impressive way, that simplified experiments in circular channels are the only justifiable approach to reduce the experimental effort and enable insights into the interlinked transfer processes by the assumption of rotation symmetry. Anyhow, the quantitative analysis of the wake structure behind rising Taylor bubbles leads to a proof of the direct link between fluid viscosity and energy dissipation inside a multi-phase system.This could be another tool to adjust the mixing time scales behind Taylor bubbles, if two-phase flow reactive systems should be investigated.
In the following, all experiments using laser induced measuring techniques on fluid motion or local mass transfer processes focus on Taylor bubbles in circular cross sections due to their higher reproducibility ensuring accurate results.

### 4.1.4 Influence of surfactants on Taylor bubble dynamics

The strong influence of the fluid properties on the local velocity fields shows, that also unwanted impurities can have a very strong effect on local processes, even if the global effects are negligible. Therefore the influence of surface active agents (surfactants) is characterized in the $D = 6$ mm channel.

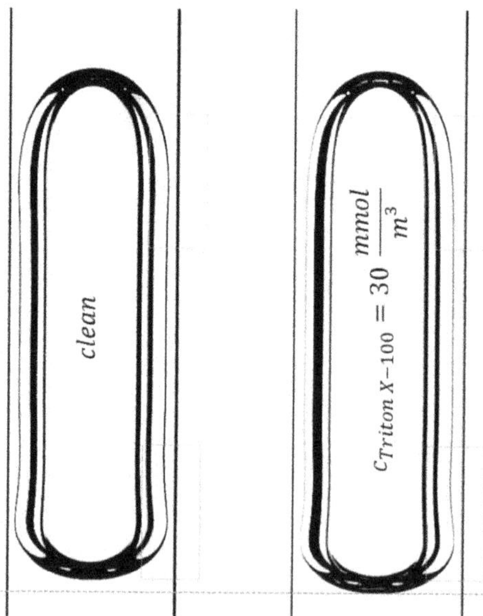


Fig. 4.22 Taylor bubble shape in clean and contaminated system in $D = 6$ mm channel.

Surfactants are influencing the bubble rise and the local liquid velocity and with increasing surfactants concentration this effects becomes stronger. But the accumulation of surfactants at the bubble interface are not only affecting the interface mobility. The strongest effect especially for Taylor bubbles with a relative low Eötvös number is due to the reduction of the interfacial tension by enriched concentration of surface active agents. An air Taylor bubble in a $D = 6$ mm channel kept in counter current flow in deionized water is shown in Fig. 4.22 left to investigate the effects of Triton X-100 on the bubble shape. Therefore, the fluid was switched from clean deionzed water to a solution with Triton ($C_{Triton} = 30$ mmol m$^{-3}$) while the bubble shape evolution to a fully contaminated state was recorded (compare Fig. 4.22 left and right). The two images show the identical gas bubble before and after contamination with Triton X100 and it is obvious that the reduction in surface tension, elongates the bubble, where the maximal diameter of the bubble is reduced, therefore the fluid film thickness is increased. As shown before in Chapter 4.1.1 slight changes in the fluid film thickness can have strong effects on the Taylor bubble rise velocity. On one hand, this lead to the expectation, that contaminated bubbles rise faster than clean Taylor bubbles, but on the other hand, it is known, that clean free rising bubbles are rising faster than contaminated free rising bubbles, due to the immobilization of the interface and the reduced internal circulation (see Chapter 2.2.1). Therefore a comparison of the rise velocity of clean and contaminated bubbles in a Triton X-100 solutions is shown in Fig. 4.23 a). Contradictory to contaminated free rising bubbles, where surfactants lower the rise velocity due to the reduced interface mobility, here the contamination results in a higher rise velocity compared to clean bubbles. Of course, surfactants are reducing the interface mobility and could lead to decreasing rise velocity even for those bubbles, but the effect is relatively small compared to the wall effects [Olg13]. A possible explanation could be, that the liquid film thickness between bubble and wall is defined by the interfacial tension (as seen in Chapter 4.1.1), where the empirical correlations are established by a function of the Eötvös number $Eo$. Therefore if the contamination leads to a reduced interfacial tension, the bubble is easier to be deformed by acting forces and this results in a thicker liquid film and larger gap area and faster rising Taylor bubble ($Eo < 10$). The effect of Triton X-100 on the global and local fluid dynamics at Taylor bubbles is investigated in the $D = 6$ mm channel. The corresponding surfactants concentration of Triton X-100 is $C_{Triton} = 10$ mmol m$^{-3}$, which is far below the critical micelle concentration of $C_{Triton,CMC} = 290$ mmol m$^{-3}$. The terminal bubble rise velocity of the Taylor bubbles are generally higher than those in clean aqueous systems in Fig. 4.23 a). But with increasing equivalent bubble diameter the velocity differences between clean and contaminated bubbles are decreasing. Actually, the volume depended rise velocity of the contaminated bubbles show that these elongated bubbles are not in the Taylor bubble regime anymore. Anyhow, the bubbles with $d_{eq} = 9$ mm show that the contaminated bubble rise nearly 50% faster as the clean bubble of the same size. To visualize the effect of Triton on the inflow velocity profile to keep a bubble of $d_{eq} = 9$ mm in counter current flow, four different Triton concentrations have been used in Fig. 4.23 b). Here it is shown, that with increasing Triton concentration the magnitude of the inflow profile increases, but a further increase over $C_{Triton} = 10$ mmol m$^{-3}$ has no measurable effect anymore. The reason could be that the surfactants concentration distribution at the lower region of the bubbles become steady and the local surface tension at the bottom is not changing anymore. Then, there is nothing forcing the bubble into another shape, and no further fluid film thickness increase occurs that would cause a higher rise velocity and a higher equivalent counter current flow. This dependency between Triton X-100 concentration and the rise velocity of elongated bubbles in channels should be checked by numerical investigations. When the surface tension dominates the flow regime, the effects of surfactants on the fluid dynamics are even more complex, because the drag coefficient is not dominated by the slip boundary condition of the interface, but a complex coupling with the fluid film thickness occurs.

In Fig. 4.24 and Fig. 4.25 average raw PIV images (averaged over five images taken at 200Hz ) and appropriate local velocity fields at a clean and a contaminated bubble front are shown.

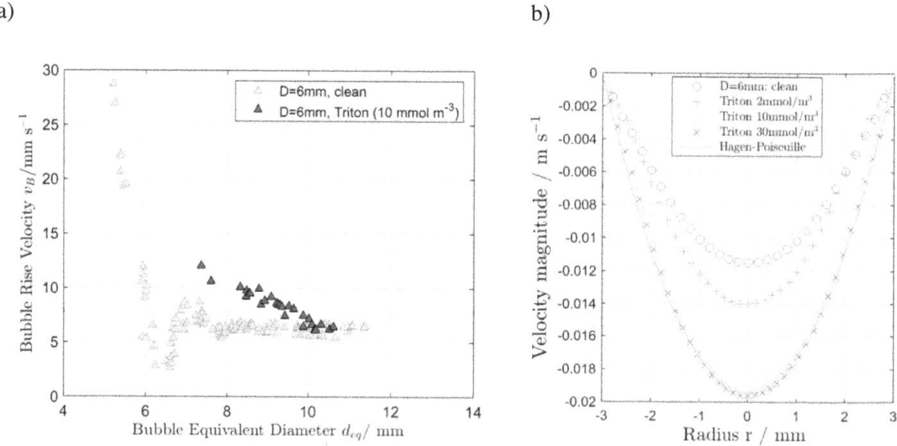


Fig. 4.23 The effect of Triton X-100 on the terminal rise velocities and local velocity profile of the equivalent counter current flow. a) The effect of Triton X-100 on the terminal rise velocities $v_B$ of Taylor bubbles in a $D = 6$ mm channel.; b) The effect of Triton X-100 on the velocity profiles in front of Taylor bubbles to keep them in counter current flow in a $D = 6$ mm channel

The average raw PIV images already indicate regions of high and low liquid velocity by the long exposure of the deposition of tracer particles. Points at the wall indicate the low velocity and long tracer lines the high fluid velocity. The velocity fields on the right side show again the fluid inflow of the counter current flow and the bubble influence due to the fluid displacement and acceleration into the film flow. The bubble front is indicated by the horizontal shadow pointing to the right side. The bright region at the bubble front is actually due to reflections from the tracer particles in-front of the bubble. In both velocity fields the stagnation point at $z^* = 0$ mm and $r = 0$ mm is visible. Along the bubble interface the particles are accelerated towards the fluid film. Due to the higher velocity magnitude in the contaminated case, the differences are not easy to spot, additionally, PIV filters the velocity regions with velocity gradients within the interrogation areas [Raf18]. Therefore, is unlikely that the PIV measurement represents single particles, which move slower along the interface due to reduced interface mobility by surfactants adsorption. To get results with even higher spatial resolution it should be considered to use particle tracking velocimetry or even spatial filter velocimetry [Hos12], [Hos17]. If the averaged raw PIV images are compared, it looks like a slightly thicker boundary layer of slow particles is generated compared to the clean bubble interface. Due to the small thickness, this cannot be resolved with PIV. Anyway, numerical investigations regarding the internal gas circulation would help to understand the local processes at the bubble interface. If the above assumptions are correct, it is possible that internal flow structures inside the bubble show different zones of gas circulation in between the front, the main bubble body and the bubble bottom. Due to the acceleration of the fluid from the stagnation point at the bubble front at $z^* = 0$ mm and $r = 0$ mm along the curved interface, the convection of a possible species transfer is increasing with increasing angle. Therefore, the concentration boundary layer could be affected locally, which would induce zones around the bubble, where the mass flux across the interface is enhanced or reduced in terms of contact time of fluid elements with the interface regarding the penetration theory (compare Chapter 2.2.2). Therefore, it is important to compare the local and global mass transfer processes at Taylor bubbles in circular channels to obtain deeper insights into the coupling of the transport processes.

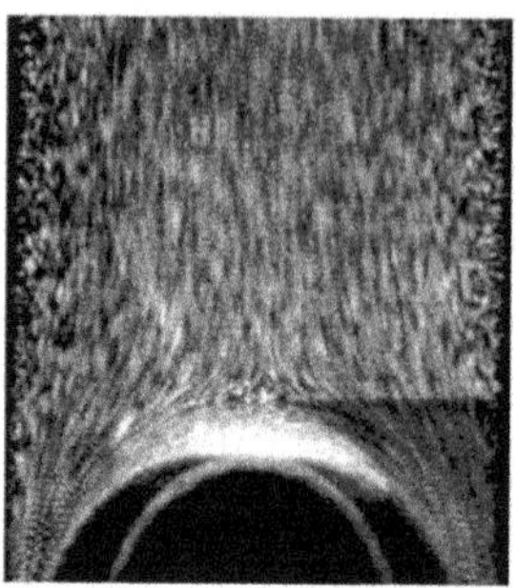

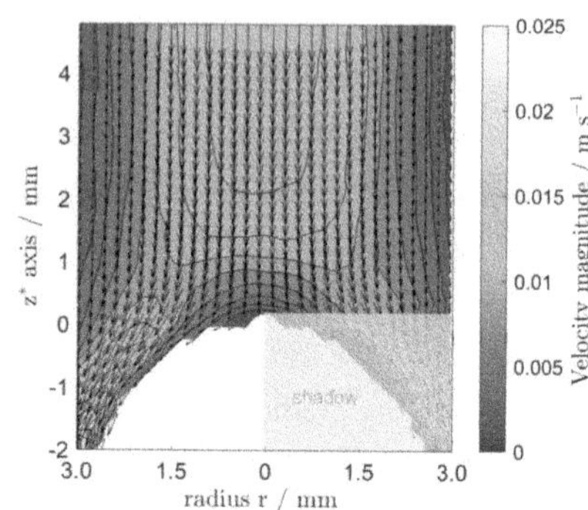


Fig. 4.24 Velocity field at a clean Taylor bubble in counter current flow in $D = 6$ mm. Left: Average PIV image (5 images; 200 Hz) to visualize streamline like structures. Right: Velocity field evaluated with PIV.

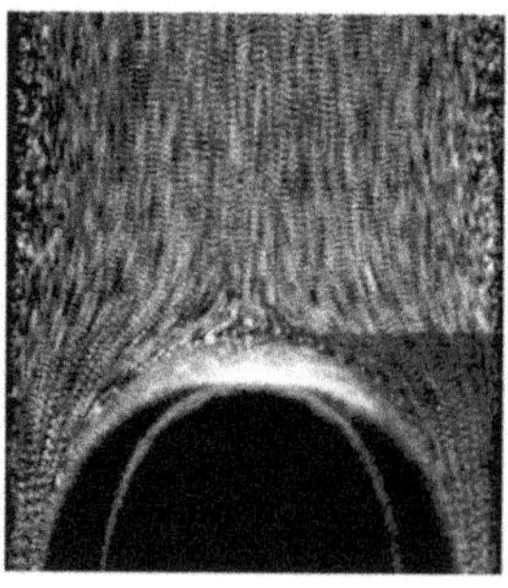

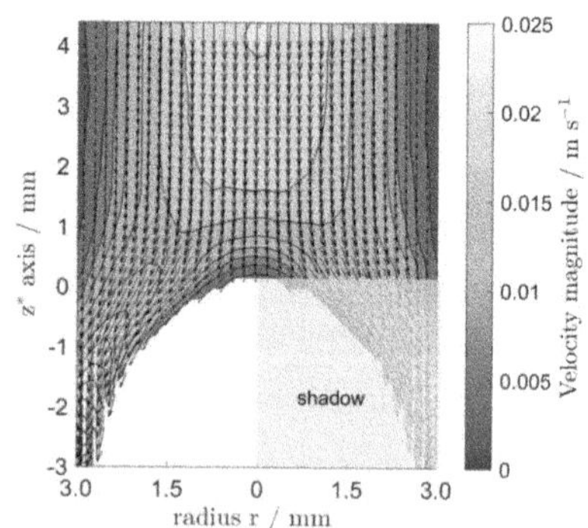


Fig. 4.25 Velocity field at a contaminated Taylor bubble in counter current flow in $D = 6$ mm. Left: Average PIV image (5 images; 200 Hz) to visualize streamline like structures in the small layer of less moving particles at the interface. Right: Velocity field evaluated with PIV.

## 4.2 Investigation of global and local mass transfer processes at Taylor bubbles

Taylor bubbles and their unique rise behavior enable detailed investigations and comparison of local and global processes through unusual long timescales for bubble dynamics and enable new insights into mass transfer processes. Therefore in this chapter, the effect of channel diameter and cross section on mass transfer are quantified via the volume reduction method Chapter 3.2.1. A Sherwood correlation for Taylor bubbles is established, enabling the prediction of long-term dissolution processes. Also, the effects of surfactants are shown exemplary and underpin that it is necessary to investigate local processes. Laser Induced Fluorescence enables the visualization of the development of a concentration boundary layer along the bubble tip. Accessible wake structures behind bubbles are characterized and local $CO_2$ concentration fields are calculated. All these insights into mass transfer processes are finally compared and quantified with a mass balance for two significant different Taylor bubble cases.

### 4.2.1 Global mass transfer processes - Influence of channel geometry

#### Mass transfer coefficient and the effect of the hydraulic channel diameter

Figure 4.26 shows the measured mass transfer coefficients $k_L$ for $CO_2$ bubbles with constant rise velocities. The $k_L$ is independent of $d_{eq}$, whereas it increases with $D_h$. It should be noted that the mass transfer coefficients for the ducts and pipes ($D_h$ = 6 and 8 mm) are almost the same. Hence the influence of the cross-sectional geometry is very small, in spite of the large difference in the rise velocity $v_B$ (see Fig. 4.23 a) ). These results indicate that the effect of cross-sectional geometries on the mass transfer from single Taylor bubbles in a mini channel is small and the mass transfer might be expressed in terms of the Sherwood number

$$Sh_D = \frac{k_L \cdot D_h}{D_L} \tag{4.2}$$

using the hydraulic diameter $D_h$ as the characteristic length and the ratio of the mass transfer coefficient $k_L$ and the diffusion $D_L$.

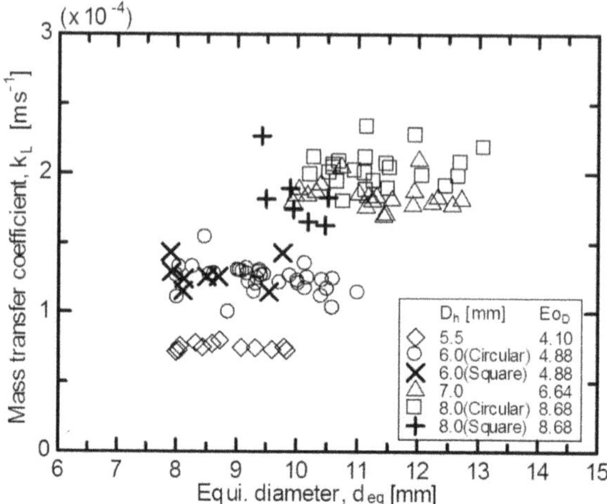


Fig. 4.26 Mass transfer coefficient $k_L$ vs. sphere volume equivalent bubble diameter $d_{eq}$ in circular and square channels 5.5 mm $< D_h <$ 8 mm [Kas15].

### 4.2.2 Modeling of global mass transfer processes by Sherwood number correlation & the prediction of long-term dissolution processes

#### Sherwood number correlation for Taylor bubbles in various channels

The measured Sherwood numbers, $Sh_D$ are plotted against $Eo_D$ as shown in Fig. 4.27. $Sh_D$ for single Taylor bubbles in larger pipes of $D_h = 12.5, 18.2$ and 25 mm quoted from Hosoda et al. [Hos14a] are also shown in the figure. As expected, all the data collapse onto a single curve, and the Sherwood numbers, $Sh_D$ can be well correlated with

$$Sh_D = 290\, Eo_D^{0.524} - \left[\frac{1.23}{(Eo_D+1)^{0.0517}}\right]^{50.1}. \tag{4.3}$$

This empirical correlation is valid for mass transfer from single $CO_2$ Taylor bubbles rising in deionized water through vertical small and intermediate channels of 5.5 mm $< D_h <$ 25 mm. The applicable ranges of dimensionless groups are $4 < Eo_D < 85$, $13 < Re(= \rho_L \cdot v_B \cdot d_{eq} \cdot \eta^{-1}) < 4.6 \cdot 10^3$, $6.2 \cdot 10^3 < Pe(= v_B \cdot d_{eq} \cdot D_L^{-1}) < 2.2 \cdot 10^6$, where $Re$ is the bubble Reynolds number and $\eta$ the liquid viscosity. Comparisons of $Sh_D$ between Eq.4.3 and the experimental data are shown in Fig. 4.28. 89% of the data are within 10 % deviations and the maximum deviation is less than 23 % [Kas15]. Judging from the analogy between the momentum and mass transfer, $Sh_D$ should be correlated in terms of $Eo_D$, since the Froude number for single Taylor bubbles is well expressed in terms of $Eo_D$ [Whi62] [Kur13]. The redefined Eötvös number takes into account the fluid property, interfacial tension, and the channel diameter. These parameters are dominating the Taylor bubble shape by the curvature of the bubble nose, the thickness of fluid falling film and therefore the interfacial area [Bro65].

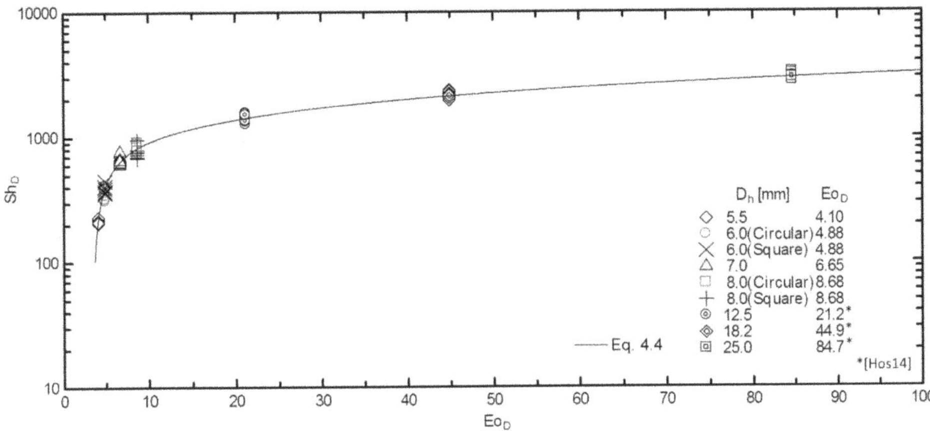


Fig. 4.27 Sherwood $Sh_d$ vs. Eötvös number $Eo$ for single $CO_2$ Taylor bubbles in various channels $5.5 < D_h < 25$ mm [Kas15].

**Prediction of Taylor bubble long-term dissolution processes**

The applicability of Eq.4.3 to a bubble dissolution process was examined through comparisons between measured and predicted long-term dissolution processes. The bubble dissolution processes were predicted using the following conservation equation of the number of moles $M_i$ of the component $i$ in the gaseous phase:

$$\frac{dM}{dt} = -k_{L,i} \cdot \pi \cdot d_{eq}^2 \cdot (C_{S,i} - C_i) \tag{4.4}$$

Since mass transfers of $N_2$ and $O_2$ are not negligible in the bubble dissolution process of a $CO_2$ bubble [Hos14a], Eq. 4.4 was solved for the three gas components, i.e. $CO_2$, $N_2$ and $O_2$. The physical properties of these gases [Him64] are summarized in Table 1.

The molar concentrations $C$ in water in this table are calculated by assuming the equilibrium with the atmosphere. The mass transfer coefficient $k_{L,i}$ was evaluated by using Eq. 4.4. Applying Henry's law to $C_S$ yields

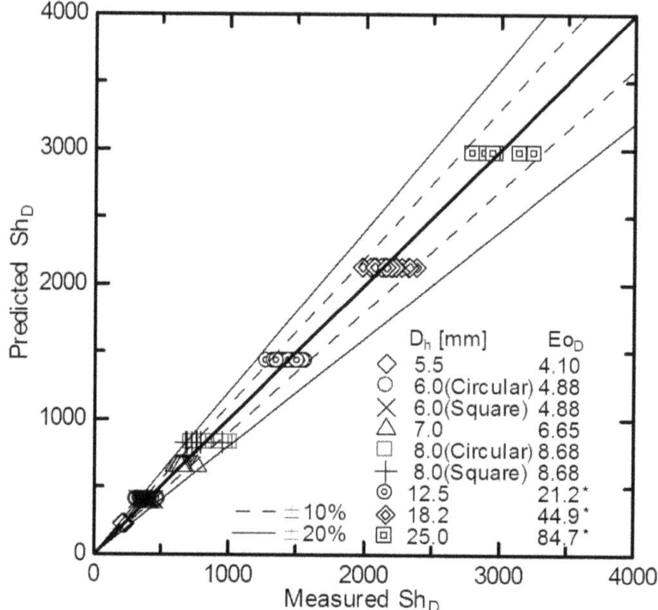


Fig. 4.28 Predicted vs. measured Sherwood number $Sh_d$ for single $CO_2$ Taylor bubbles in various channels $5.5 < D_h < 25$ mm [Kas15].

Table 4.1 Physical properties of gases in water at T = 298 K [Him64]

| | $CO_2$ | $N_2$ | $O_2$ |
|---|---|---|---|
| $H$ [GPa] | 0.166 | 8.55 | 4.41 |
| $D_L$ [m$^2$ s$^{-1}$] | $1.9 \cdot 10^{-9}$ | $2 \cdot 10^{-9}$ | $2.3 \cdot 10^{-9}$ |
| $C$ [mol m$^{-3}$] | 0.011 | 0.51 | 0.27 |

$$C_{S,i} = \frac{C_V \cdot p \cdot X_i}{H_i} \tag{4.5}$$

here the mole fraction $X_i$ of the component $i$ in the gaseous phase is given by

$$X_i = M_i / \sum_{i=1}^{N_m} M_i \tag{4.6}$$

where $N_m$ is the total number of gas components, i.e. $N_m = 3$. The bubble equivalent diameter can be expressed in terms of $M_i$ by using the equation of state for the ideal gas:

$$d_{eq} = \left[ \frac{6}{\pi} \left( \sum_{i=1}^{N_m} M_i \cdot R \cdot T_L / p \right) \right]^{\frac{1}{3}} \tag{4.7}$$

Fig. 4.29 shows comparisons between measured and predicted long-term dissolution processes in circular channels with a diameter of $D_h = 6, 8$ and 12.5 mm. The initial bubble diameter is $d_{eq,in} = 12.4$ mm for $D_h = 6$ mm, $d_{eq,in} = 15.5$ mm for $D_h = 8$ mm and $d_{eq,in} = 26.7$ mm for $D_h = 12.5$ mm [Hos14a], respectively. The initial mole fraction are $X_{CO_2}(0) = 0.8$ and 0.5. The time $\tau$ in the figure shows the time elapsed from the bubble arrival at the measuring section. The diameter decreases due to dissolution, and becomes almost constant, which means that the mass transfer process has reached an equilibrium state. The bubbles in the equilibrium state are still Taylor bubbles due to low initial $CO_2$ mole fraction, which results in a larger remaining bubble as for initial pure $CO_2$ bubbles. The predictions agreed well with the experiments for all the cases. This means that Eq. 4.4 is applicable to predictions of long-term bubble dissolution processes. The prediction of long-term bubble dissolution processes for Taylor bubbles in square channels is possible with this equation as well, but the results are not proposed in this work, because of the reduced reproducibility of the experiments [Kas15].

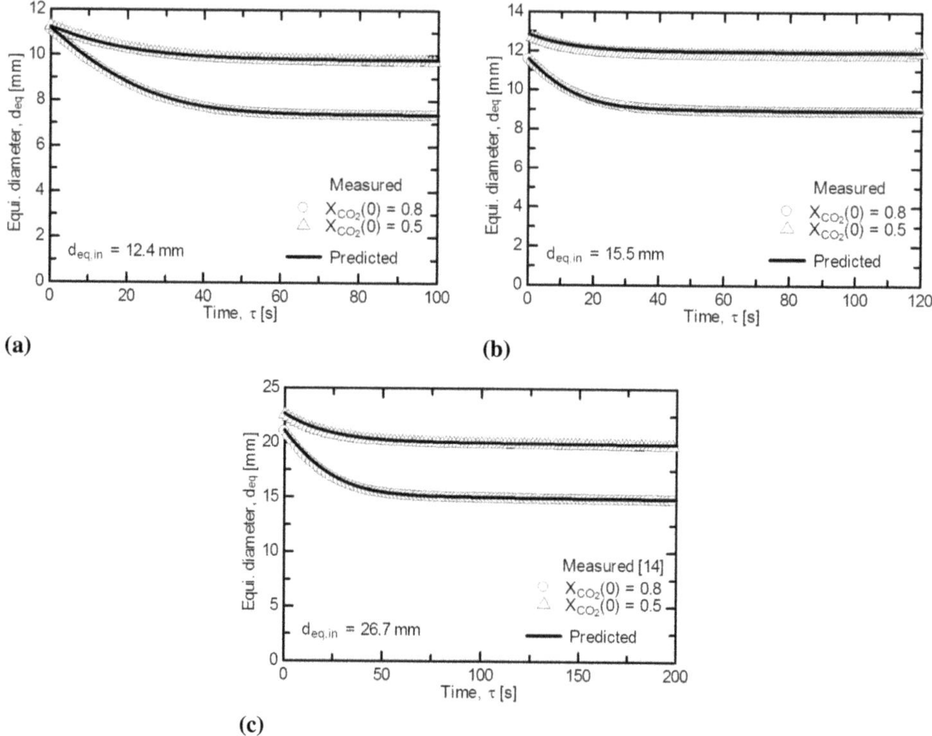


Fig. 4.29 Comparison of measured and predicted long-term dissolution process of $CO_2$ Taylor bubbles $d_{eq}$ over time in a) $D = 6$ mm, b) $D = 7$ mm and c) $D = 8$ mm. Initial equivalent bubble diameters are $d_{eq,in} = 12.4, 15.5$ and $26.7$ mm [Kas15].

### 4.2.3 Influence of surfactants on Taylor bubble mass transfer

#### Triton X-100 and the influence on global mass transfer coefficient

The effect of Triton X-100 on $CO_2$ Taylor bubbles in channels $D > 12.5$ mm has been already shown by Hosoda et al. [Hos14a] indicating that the rise velocity is not affected but the mass transfer coefficient is reduced significantly. In comparison, the effect the identical concentration of Triton X-100 on the fluid dynamics on Taylor bubbles in smaller channels ($D = 6$ mm) shows that surfactants might have a strong effect on systems with low $Eo < 10$, where the elongated bubbles actually rise faster than clean bubbles, see Chapter 4.1.4. PIV measurements indicated a reduced tangential convective fluid transfer directly at the bubble front (Fig. 4.25). This might increase the contact times of the fluid elements at the bubble tip, which could cause a reduced local and overall mass transfer performance, while the fluid elements flow down the interface. Even though the terminal rise velocity is higher for the contaminated compared to clean bubble Fig. 4.23 and Fig. 4.25 most likely due to a thicker fluid film between bubble and wall.

In Fig. 4.30 the mass transfer coefficient $k_L$ over the equivalent bubble diameter $d_{eq}$ for clean (bright triangles) and for contaminated (dark triangles) Taylor bubbles in a $D = 6$ mm channel are shown. The used surfactants concentration is $C_{Triton} = 10$ mmol m$^{-1}$, the same concentration as used before. As expected, Triton reduces the overall mass transfer coefficient $k_L$ strongly about a factor of 0.5 compared to the clean $CO_2$ Taylor bubbles in the same channel. The reason of the mass transfer reduction is explained by the additional resistance of the molecular layer of the adsorbed surfactants. This finding confirms the observed local fluid dynamic phenomena at the interface, even if the overall rise velocity would suggest to find the opposite trend, due to the Sherwood correlations based on *Re*. This shows that local transport processes at the interface are dominant and overall measured values like rise velocities can miss out significant trends and effects. The contradictive trends for the overall rise velocity and mass transfer for contaminated systems, might be explained by the part of the bubble, which dominates the effect. While the rise velocity is dominated by the shape of the bubble bottom, the mass transfer is dominated by the transfer at the tip of the bubble that gets in contact with the fresh liquid phase resulting in a high driving force $\Delta C$.

### 4.2.4 Local mass transfer investigations by means of Laser Induced Fluorescence

The comparison of global and local transport processes at Taylor bubbles in vertical channels have already shown that local investigations are needed to understand the mass transfer

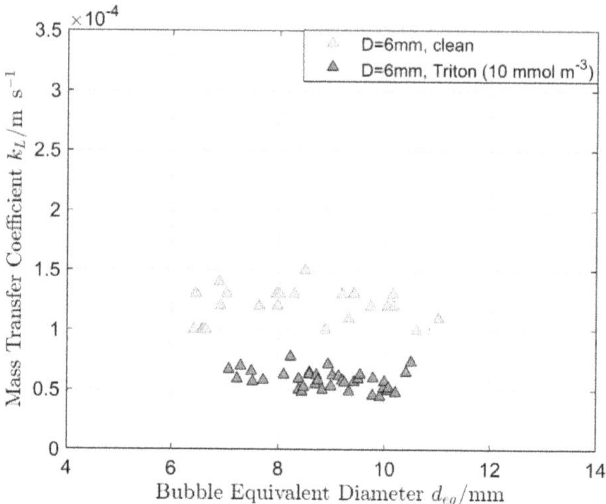


Fig. 4.30 Influence of Triton X-100 ($C_{Triton}$ =10 mmol m$^{-3}$) on the mass transfer coefficient $k_L$ of $CO_2$ Taylor bubbles in a $D = 6$ mm channel.

processes across interfaces. Additionally, the transport processes at Taylor bubbles are different as expected from literature of free rising bubbles. For example the finding of a stagnation point in the flow field at the Taylor bubble tip requires local investigations of the concentration boundary layer. Therefore, Laser Induced Fluorescence as a non-invasive method [1] is used to estimate the concentration boundary layer thickness $\delta_c$ at the Taylor bubble tip, to visualize the wake structure and to estimate the concentration field and characterize the mixing behind bubbles in vertical channels.

### $CO_2$ boundary layer and local Sherwood number at the bubble front

The high controlability of Taylor bubbles kept in counter current flow in vertical channels and their steady interface are ideal for local mass transfer investigation, even for several seconds, which is an untypical long observation period for bubbly flows. As shown before in Chapter 4.1.3, the local fluid velocity along the interface at the Taylor bubble front is increasing, which should enhance the tangential convective transfer, which has to be confirmed or confute by local investigations on mass transfer processes.
In Fig. 4.31 (left) and LIF image of the tip of a $CO_2$ Taylor bubble in a $D = 6$ mm channel is

[1] Because fluorescent dyes can be surface active it is necessary to check in advanced the effects on the terminal rise velocity and overall mass transfer coefficient to reduce the risk of contamination effects of the dye on the local transport processes, which should be measured. In appendix D.1 a) and b) it is shown, that the LIF with fluorescein sodium salt $C_{dye} = 10$ mmol L$^{-1}$ as fluorescent dye can be seen as non-invasive.

shown. The laser is introduced from the right side and generates a bubble shadow to the left. The red rectangle is indicating the Region Of Interest. The dark thin line from the upper left to the lower right in the ROI is the concentration boundary layer due to reduced fluorescence signal of the dye, caused by the local pH reduction by dissolved $CO_2$. This boundary layer thickness is measured by the distance between two edge detected lines of the boundary layer using Matlab image processing as seen in Fig. 4.32 (right), where horizontal and vertical pixelwise distances are plotted to show the actual evolution along the interface. It can be seen, that the concentration boundary layer thickness $\delta_c$ is decreasing with increasing angle $\alpha$ or radius $r$ as expected from the velocity field at Taylor bubble tips shown in Fig. 4.10 or 4.14. The tangential convective transport is increased along the interface, and at a certain distance to the interface, the convective mass transfer in tangential direction dominates the diffusive mass flux in normal direction. The visible edge between the dark boundary layer and the bright bulk in the grayscale values of the raw LIF image indicates therefore the position, where equilibrium between the diffusive and the convective mass transfer is reached. With

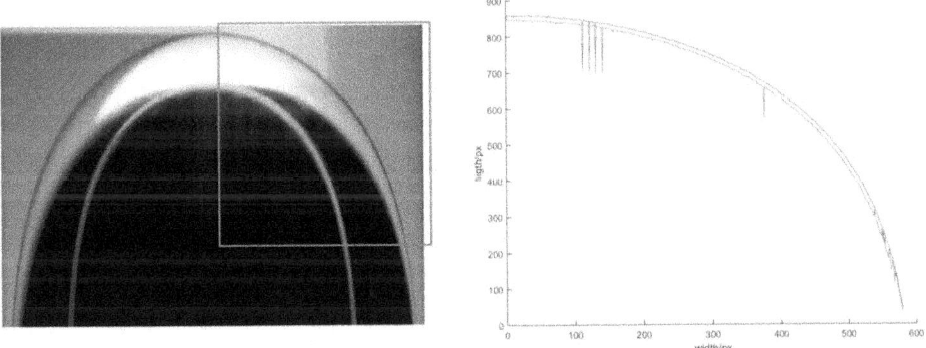


Fig. 4.31 Evolution of the concentration boundary layer $\delta_c$ at a Taylor bubble front kept in counter current flow in a $D = 6$ mm channel. Left: LIF image of a Taylor bubble front with significant dark boundary layer, which becomes thinner with increasing $r$ due to enhanced local tangential convection. Right: Evaluation method for $\delta_c$ via edge detection.

this result a rough estimation of the local Sherwood number is possible by estimating the horizontal and vertical distance $\delta_c$ of the two lines. To achieve more accurate data was not possible in the current set-up due to comparable low spatial resolution of this multiscale problem (e.g. $D = 6$ mm vs. $\delta_c < 100\ \mu$m) and the chip noise. In any case, the local Sherwood number was estimated to be $Sh_{loc}(r = 0\text{ mm}) = 50$ and increases exponentially with the angle $\alpha$ along the bubble interface, due to the fluid acceleration. The scale of the local Sherwood number seems to be reasonable, even though the exact value is difficult to compare with literature values of freely rising bubbles due to the differences in fluid flow and the unknown internal gas circulation zones. The conditions and values are similar to recent experiments by Rüttinger et al. [Rüt18] and even typical overall Sherwood correlations (Eq. 2.34) lead to in the similar ranged values $Sh \approx 50$, if the bubble Reynolds number is used, which is not taking into account local phenomena at the interface.

But it is clearly visible, that the evolution of the concentration boundary layer is very different to free rising bubbles and $\delta_c$ becomes too small to be measured quickly, due to the fluids acceleration as shown in Fig. 4.32 b). The local Sherwood number $Sh_{loc.}$ is calculated from the local concentration boundary layer thickness $Sh_{loc.,LIF}$ and compared with the local Sherwood number calculated by the local velocity magnitude at the position of the concentration boundary layer $\delta_c$ according to the empirical correlations $Sh = Re^n \cdot Sc^m$. For more reliable values a higher spatial resolution or a higher magnification is required, which makes it difficult to take an appropriate target scale. This is the reason, why the concentration boundary layer was only measured in a $D = 6$ mm channel, with its moderate local fluid velocities. Compare therefore, the Fig. 4.31 b) where the concentration boundary layer thickness evaluates from $\delta_c = 12$ pixels at $\alpha = 0°$ to $\delta_c \approx 3$ pixels at $\alpha \approx 70°$. Also a calculation of the local dissolved $CO_2$ concentration in the boundary layer was not possible and due to reflections at the interface, minor vibrations, interface motion or camera noise.

The local Sherwood number evolution around spherical bubbles is obviously very different to the development of the local Sherwood number at Taylor bubbles in Fig. 4.32. With increasing angle $\alpha$ the opposite trend is known for spherical bubbles, where $Sh_{loc}$ decreases. For spheres in open fluid the local concentration boundary layer thickness increases with the angle $\alpha$, and therefore the local Sherwood number decreases [Bra71]. In the case of a Taylor bubble the distance of lines of constant concentration in the fluid become smaller, which increases the concentration gradient due to the wall effect and the accelerated fluid motion in the film region. On one hand, according to the Two Film Theory the local mass transfer at $\alpha = 70$ is two times higher than at the Taylor bubble tip $\alpha = 0$, because the hydrodynamic boundary layer $\delta_f$ and the concentration boundary layer $\delta_c$ are reduced by the acceleration of the fluid into the film region. On the other hand, $Sh_{loc.}$ should be reduced from $\alpha = 0$ to $\alpha = 70$ due to the Surface Renewal Theory and the contact time $\tau$ of fluid elements. This could be compensated by the increasing circumference of the bubble, which would lead to a stretching of the existing fluid elements or a refill by fresh fluid elements from the bulk (convectional transfer). This shows again even simplified mass transfer processes are complex 3D interlinked transport processes

These experiments have shown, that the evolution of the boundary $\delta_c$ can be measured qualitatively at the interface with LIF, whereas the estimation of the local dissolved amount of

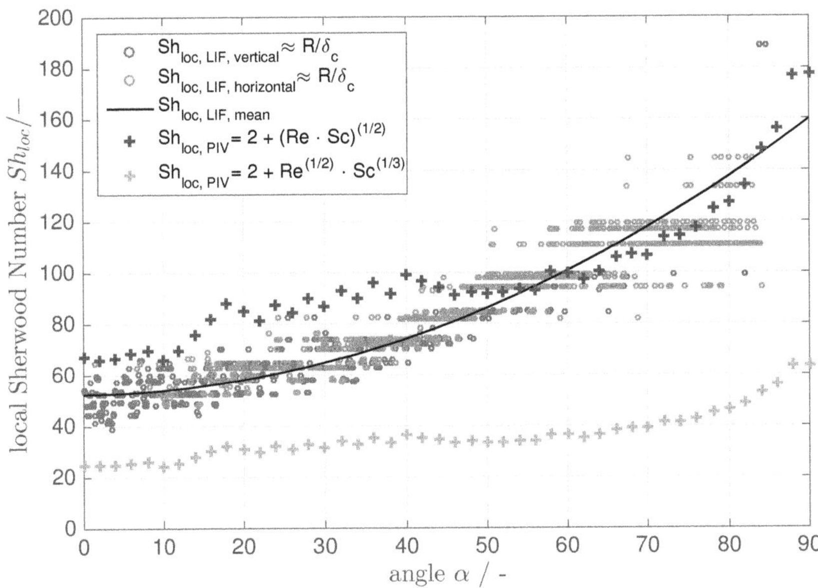


Fig. 4.32 Estimation of the local Sherwood number $Sh_{loc}$ from LIF and PIV data at Taylor bubble front between angle $\alpha = 0°$ and $90°$. The relation of the channel radius $R$ and $\delta_c$ taken from LIF images, compared with $Sh_{loc}$ calculated from local velocity field (PIV), where the Levich relation $\delta_c/\delta_{hyd} = Sc^{1/3}$ is used to choose the local velocity values to calculate Froessling correlation e.g. $Sh_{loc} = 2 + (Re^{1/2} \cdot Sc^{1/2})$

$CO_2$ is not possible directly at the interface. Additionally, the direct connection between the increasing tangential convection and the decreasing concentration boundary layer was shown and even plausible values of $Sh_{loc}$ could be derived along the interface. Due to the elongated Taylor bubble shape it is not practicable to measure local dissolved $CO_2$ at the interface and in the wake structure due to the multiscale problem, where the Taylor bubble plus the wake region is quickly in the range of $\Delta z > 10$ D, which makes it impossible to resolve both at the same time. In the following, the focus is on the transferred and dissolved $CO_2$, which is mixed in the wake region with the bulk of the liquid phase, where the calibration is applicable and the local amount of $CO_2$ can be measured and visualized.

**Concentration wake structures behind Taylor bubbles - Influence of channel diameter and dissolution time**

$CO_2$ is transferred across the interface and transported downwards from the concentration boundary layer into the wake region of the bubbles, where it is mixed with the bulk of the fluid phase. Here the calculation of the local dissolved $CO_2$ is suitable via a pixelwise calibration matrix. In Fig. 4.33 the concentration fields behind three bubbles during a long-term dissolution of a Taylor bubble in counter current flow in a $D = 7$ mm channel is shown. Fig. 4.33 a) shows the LIF image of the Taylor bubble with its wake. The automated detection of the bubble interface for masking the bubble in this figures proved to be involved. The reason is the non-uniform gray values around the interface due to reflections, shadows and the induced fluorescence by the dye. Therefore, the evaluation of dissolved $CO_2$ inside the bubble should be ignored in the following.

On the right side, the local dissolved $CO_2$ concentration field at $t_1$ is shown. $CO_2$ is accumulated in the toroidal vortex, known from the PIV measurements in Fig. 4.16. During long-term bubble dissolution the mole fraction of $CO_2$ inside is shifting, while $N_2$ and $O_2$ concentration increases, as shown before in Fig. 2.15 by Hayashi et al. Therefore, the mass transfer is reduced and lower local concentrations occur inside the wake region, which is visible in Fig. 4.33 b) at $t = t_2 = t_1 + 10$ s. Note that for the intermediate bubble size (compare Fig. 4.1) the global and local dynamics enhance, indicated by higher vortex shedding in the wake. This development is continued until the bubble is small at $t = t_3 = t_2 + 20$ s in Fig. 4.33 c), with even lower $CO_2$ concentrations and higher vortex shedding in the wake. This shows that even long-term transport processes about more than 30 seconds can be visualized and quantified with Laser Induced Fluorescence at Taylor bubbles.

Note that in all three concentration fields the $CO_2$ concentrations lower than $C_{CO2} = 0.05$ mol $L^{-1}$ are blanked for a better visibility. This hides some low concentration regions in small vortices but helps to understand the important regions of bulk and wake. These insights into local mass transfer processes in two-phase flows need to be quantified and validated with overall mass transfer measurements.

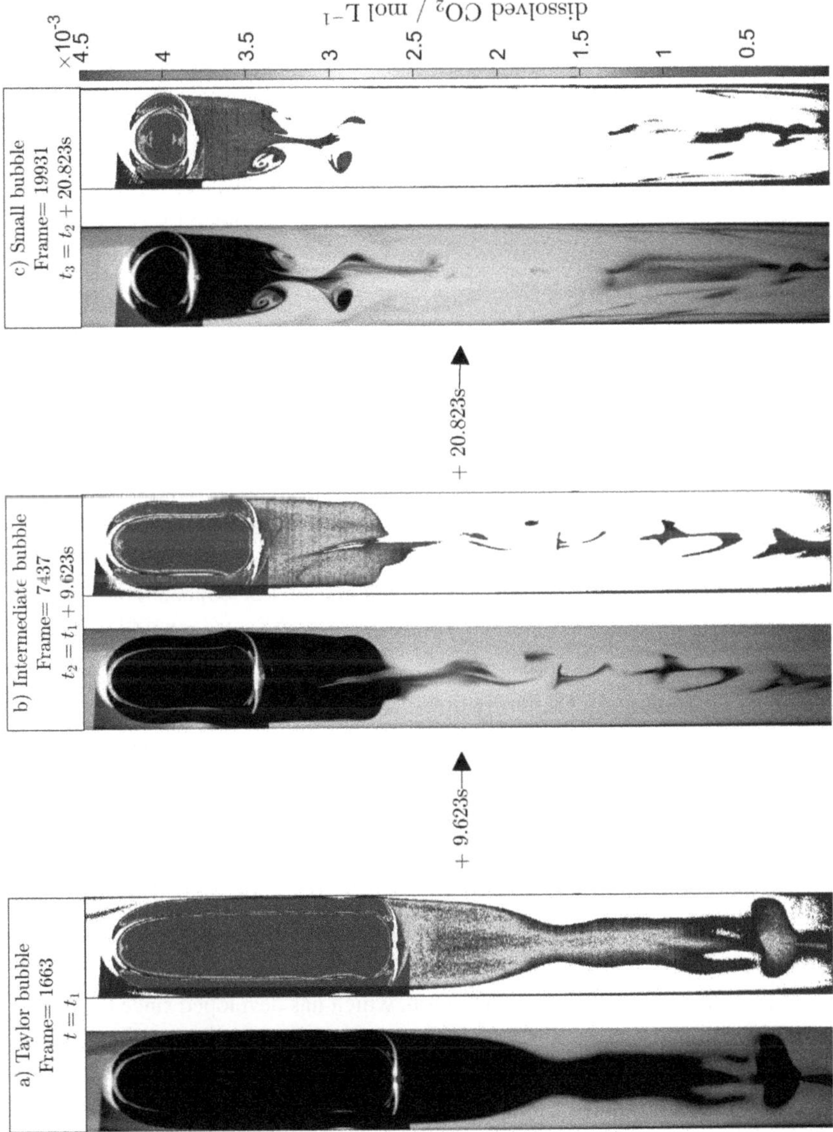


Fig. 4.33 Time evolution of a concentration field during a long-term dissolution of a Taylor bubble kept in counter current flow in a $D = 7$ mm channel. The local $CO_2$ concentration in the wake is reduced, while the mole fraction shift inside the bubble, due to counter diffusion of predissolved $O_2$ & $N_2$.

## 4.3 Comparison of global and local measuring techniques - Bubble volume difference vs. the integration of concentration field

The extraordinary properties of Taylor bubbles in vertical circular channels for optical measuring techniques, desire a combined approach of global and local imaging methods for a mass balance of the transferred gas specie. There is no other two-phase experiment which combines so many advantages like symmetry, low rise velocity and therefore enables high spatial and temporal resolution, easy optical access and steady interfaces while using no model fluids. Due to the high mass transfer of $CO_2$ in deionized water and therefore a comparable high volume reduction of the bubble, the transferred moles of $CO_2$ can be measured (Chapter 3.2.1) to calculate accurate mass transfer coefficients $k_L$ as presented in Chapter 4.2.1. With the LIF methods and an accurate pixelwise calibration matrix, local dissolved $CO_2$ concentrations in the fluid can be evaluated as shown in Fig. 4.22 . This enables the combination of both techniques for validation deriving a $CO_2$ mass balance as presented in this chapter.

In the following a mass balances will be made for two rising Taylor bubbles in a $D = 6$ mm and a $D = 7$ mm circular channel, which have the advantages in their moderate fluid velocity and symmetrical wake structure for integral calculations of concentration fields behind bubbles. The Field of View was chosen to be about $z \approx 8 \cdot D$ to show the rising Taylor bubble and its wake at two different time steps with a significant bubble volume reduction. The total amount of transported $CO_2$ during time interval $\Delta t = t_2 - t_1$ into the fluid is estimated inside the wake, using the pixelwise calibration function. Using this advanced calibration, 2D dissolved $CO_2$ concentration fields are calculated, which are further rotated and integrated to estimate the moles of dissolved $CO_2$ (compare Chapter 3.2.3). In Fig. 4.34 a) and b) the identical Taylor bubble at two different time steps $t_1$ and $t_2$ is shown with the equivalent moles of $CO_2$ noted in the boxes above: $M_1 = 1.84 \cdot 10^{-5}$ mol, $M_2 = 1.62 \cdot 10^{-5}$ mol. Due to the time shift, the bubble rose in the channel and the bubble volume has decreased due to the mass transfer into the continuous phase, noted as mole difference $\Delta M_{2,1} = 2.20 \cdot 10^{-6}$ mol (bubble shrinkage). The dark wake region, which has developed since the position at $t_1$ of the Taylor bubble is indicated by a red rectangle at $t_2$ in b) and is cropped and magnified in c). In subplot Fig. 4.34 d) the concentration field behind the Taylor bubble in a $D = 6$ mm with $M_C = 3.49 \cdot 10^{-6}$ mol determined using the pixelwise calibration functions is shown. The ratio of transferred $CO_2$ from the bubble reduction measurement and from the $CO_2$

concentration in the wake shows a significant imbalance of $r = 0.6$. The imbalance can be

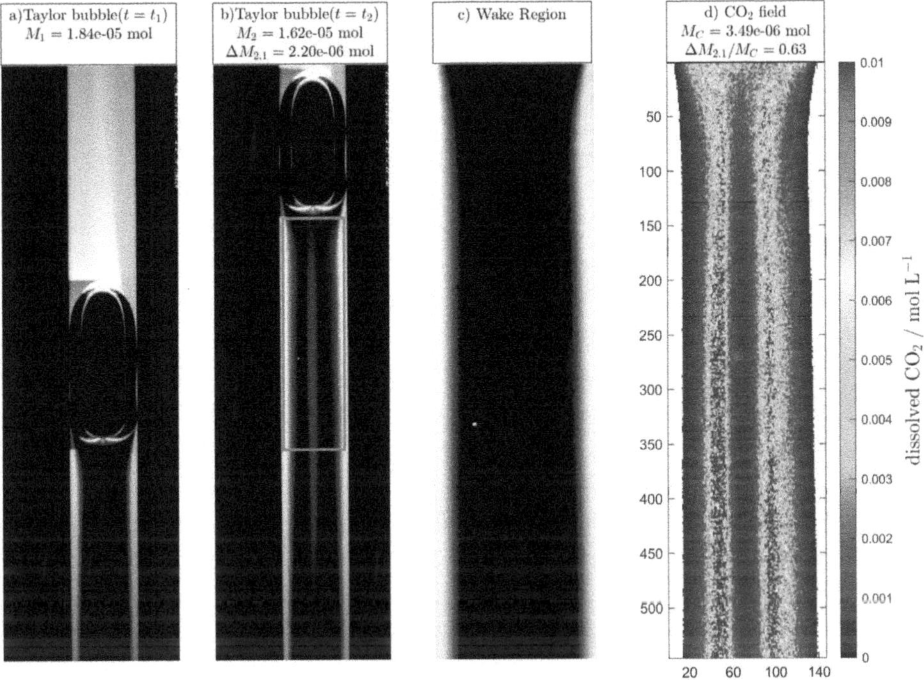


Fig. 4.34 Evaluation of a mass balance of the moles of $CO_2$ between the integration of a concentration field $M_C$ behind a rising Taylor bubble in a $D = 6$ mm, compared with the $CO_2$ mole difference $\Delta M_{2,1}$ of the bubble.

explained by the comparable large time difference $\Delta t_{2,1} < 3$ s between both images, where also $N_2$ and $O_2$ is transferred into the bubble from the surrounding liquid and reduces the volume reduction as shown in Chapter 4.2.2 or Fig. 2.15. Therefore, it is highly expected, that there are more moles of dissolved $CO_2$ than expected by the volume reduction in mean time. Although, the reduction of the local $CO_2$ concentration along the wake from bottom to top could be explained by the reduced transfer of into the fluid, as explained before. Experiments with degassed fluids would not show such a behavior and a comparison would be the next logical step.

In Fig. 2.15 the long-term dissolution process of Taylor bubbles in larger channels $D = 12.5$ and 25 mm and the simulated mole fraction is shown, to compare the dissolution with and without counter mass transfer from $N_2$ and $O_2$ [Hos14a]. After about 25 seconds of dissolution the development shows significant differences, where the gas mole fraction inside the bubble changes as well. In the current experiments the progress is much faster,

because Taylor bubbles in $D = 6$ mm channel has much higher ratio of interfacial area to its volume, than Taylor bubbles in $D = 12.5$ - 25 mm channels. Additionally, mass is transferred during the time between injection and observation, which cannot be taken into account. From the investigation of the mass transfer coefficient from bubble volume shrinking, it is concluded that the dissolution observation should be stopped as soon as after bubble injection for best accuracy or the other transferred gas species should be taken into account [Kas15]. Longer observation times for the estimation of the mass transfer coefficient have reduced the accuracy, due to a reduced volume shrinkage because of the counter mass transfer, as can be seen by the low ratio of transferred $CO_2$ to $CO_2$ in the concentration field.
These errors should be less dominant, if a Taylor bubble is rising inside a $D = 7$ mm channel, where the rise velocity is higher, which results in a smaller time difference $\Delta t_{2,1} \approx 1$ s in between the considered images for the bubble shrinkage. Further, the wake structure is still nearly rotation symmetric. Of course, the quantitative evaluation of 2D LIF requires an accurate of rotational symmetry. In Fig. 4.35 a) and b) the identical Taylor bubble at two different time steps $t_1$ and $t_2$ is shown with the equivalent moles of $CO_2$ in the boxes above: $M_1 = 3.56 \cdot 10^{-5}$ mol, $M_2 = 3.29 \cdot 10^{-5}$ mol. Due to the time shift, the bubble rose in the channel and the bubble volume has decreased due to the mass transfer into the continuous phase, noted as $\Delta M_{2,1} = 2.73 \cdot 10^{-6}$ mol. The dark wake region c), has been established since the position at $t_1$ of the Taylor bubble (red rectangle), which is further evaluated. In subplot d) in Fig. 4.35 the evaluated concentration field behind a Taylor bubble in a $D = 7$ mm channel is shown, where the integration of the total amount of $CO_2$ results in $M_C = 2.84 \cdot 10^{-6}$ mol. The ratio of transferred $CO_2$ and $CO_2$ in the concentration field shows a very good agreement for this condition, where the time difference $t = t_2 - t_1$ is comparable small resulting in less counter mass transfer of predissolved gases. Note that all grayscale images have actually a bit depth of 16bit which is converted to 8bit images to enable a suitable visualization. Additionally, in the evaluated concentration field images, regions with very low concentrations ($C < 0.05$ mol L$^{-1}$) are blanked to be marked as white instead blue as suggested from the shown color-map. This helps to show the similarity in wake shape between grayscale images and evaluated concentration fields and is not affecting the calculations.
The validation of the global and local mass transfer via two imaging techniques is a very successful approach for detailed mass transfer investigations with high spatial resolution of such complex coupled transport processes at fluid-gas interfaces. The combination of shadowgraphy and LIF to validate the mass transfer measuring methods and enable a mass balance is very promising, even for reactive multiphase flows. A transfer of this technique for investigations in other solvents are very promising as well, as the development to a two dye LIF technique. This might enable to evaluate the concentration field of more than one gas

specie inside the continuous fluid, or even, e.g. the generation of a product and side product of a competitive consecutive reaction.

## 4.4 Influence of mixing on mass transfer and chemical reaction

Many advantages of Taylor bubbles in vertical channels have been demonstrated during the investigations presented in the previous chapters. Additionally, Taylor bubbles have been chosen to be the ideal two-phase flow test system for reactive multiphase flows by the priority program of the German Research Foundation DFG SPP1740 "Reactive Bubbly Flows", where a reactant is transferred into the continuous fluid solvent. The new insights into the

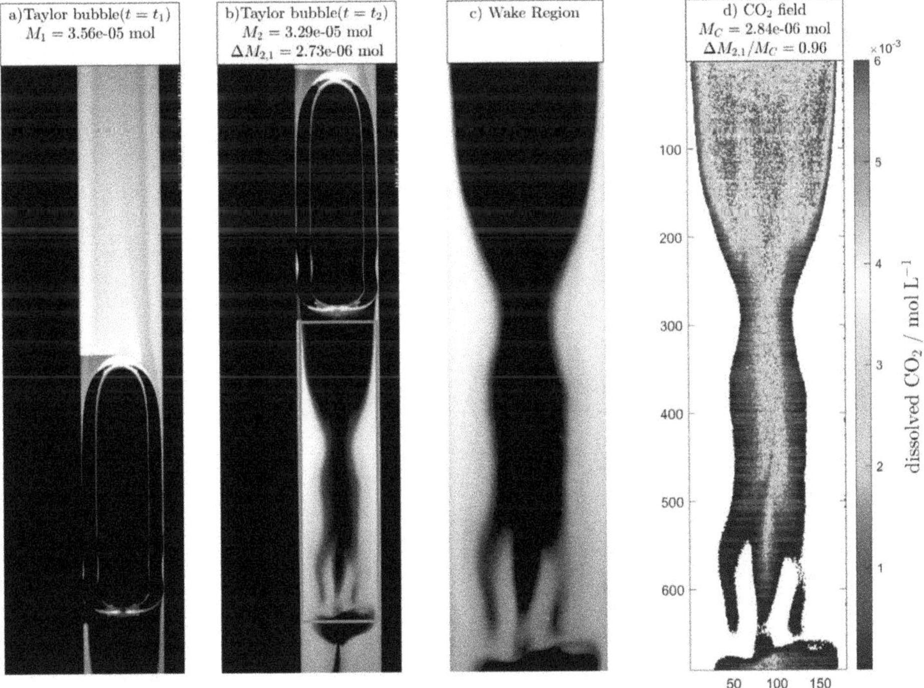


Fig. 4.35 Evaluation of a mass balance between the integrated moles of $CO_2$ of the concentration field $M_C$ behind a rising Taylor bubble in a $D = 7$ mm channel, compared with the bubble volume difference $\Delta M_{2,1}$ by shrinkage during dissolution.

transfer processes from this work have been used to build a similar experimental setup, which is solvent resistant, like for acetonitrile or methanol, and sealed against atmospheric gaseous species as reported by Kastens et. al [Kas17b]. This research has been done in close cooperation with Dr. Jens Timmermann [Tim18] and Dr. Alexandra von Kameke [vK19]. The main goal of this project is to understand the coupling of local fluid dynamics and mixing on consecutive competitive reactions, where e.g. the transferred gas specie reacts to a certain product and possibly to a consecutive side product like for an oxygenation. For this specific problem, collaborating chemist inside the project DFG SPP1740 established reactive systems with moderate reaction kinetics, which enables local investigations via color change, UV-Vis or LIF, without any dye [Str16][Str17][Pau18][Tim18][Ben18], examples are given in Fig. 4.36. With the application of these customized reactive systems the advantage of Taylor bubbles in combination with the accurate global and local measuring techniques could provide new insights into the relation of mixing and reactive timescales. Timescales and residence times in reactive multiphase flows can be essential for product selectivity. For instance when a gas species is consumed, e.g., by a competitive consecutive reaction when reaction timescales are comparable to mixing timescales. To point out the importance of the fluid flow, the analysis of the experimental velocity data from Taylor bubble wakes in a $D = 6$, 7 & 8 mm channel, as seen in Fig. 4.15 was analyzed by means of Lagrangian methods [vK19].

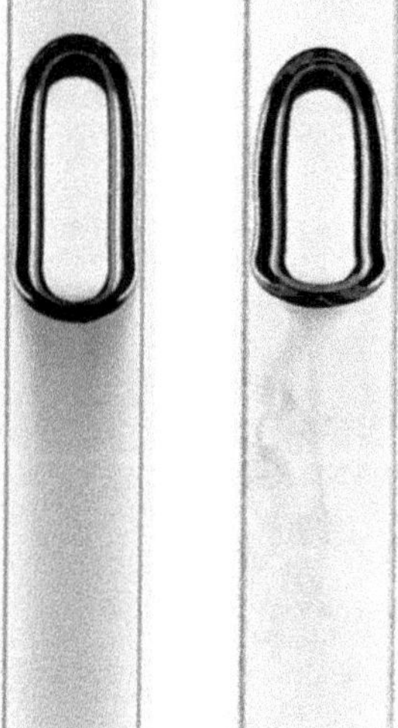

Fig. 4.36 Color change of [Cu(btmgp)I] complex due to oxygenation reaction to copper(III) (orange) and the decay to copper(II) (green) in the wake of Taylor bubbles ($D = 4$ & 5 mm; $Re = 19$ & 141; $Eo = 4.4$ & 6.8). [Pau18, Tim18]

# Chapter 5

# Conclusion

In this work the global and local fluid dynamics and mass transfer are investigated at Taylor bubbles in clean and contaminated systems to provide a deeper knowledge and understanding of the coupling of both processes. For the detailed investigations Taylor bubbles in vertical channels $5.5 < D < 8$mm are used because they overcome the problem of dynamic shape deformation and complex 3D rise trajectories. Additionally, elongated Taylor bubbles have a volume independent rise velocity and can therefore be kept in counter current flow easily, which make them accessible for detailed optical investigations with a high temporal and spatial resolution even for long-term measurements. The measuring techniques for the global and local variables are high-speed Shadowgraphy, 2D2C Particle Image Velocimetry and p-2D Laser Induced Fluorescence to observe the global terminal rise velocity and shrinking behavior of such bubbles, the local velocity fields and local concentration fields and interface shape. The comparison of local and global techniques has been used to show the accuracy of the results.

The main findings of this work are the characterization of small, intermediate and Taylor bubbles in a small $D = 7$mm channel, the bubbles local maximal radius and therefore local minimal fluid film thickness, which seems to be the major aspect that determines the rise velocity (Chapter 4.1.1). This finding explains the rise velocity jump with increasing bubble volume. Additionally, a Sherwood $Sh_D$ correlation has been developed to describe the mass transfer of $CO_2$ Taylor bubbles over a wide range of channel diameters $D = 5.5 - 25$ mm, as shown (Chapter 4.2). The mass transfer from single $CO_2$ Taylor bubbles in vertical mini-channels was measured for various channel hydraulic diameters $D_h$. Sherwood numbers, $Sh_D$, using $D_h$ as characteristic length, are well-correlated in terms of the Eötvös number $Eo$ for $CO_2$ Taylor bubbles in $5.5 < D < 25$ mm channels. The proposed $Sh_D$ correlation is in good agreement with experimental results for a long-term dissolution process of Taylor

bubbles (Fig. 4.27).
The identification of significant different wake structures behind Taylor bubbles in channels $D = 6,7$ and 8mm, like laminar, toroidal and turbulent mixing, via high-speed 2D2C-PIV have become another advantage of the unique two-phase flow system. The ability to transfer the flow characteristics to solvents like acetonitrile or methanol keeping $Eo$ constant, made Taylor bubbles to a promising tool for reactive multiphase flows.

Another main finding, is the measured reduction of the $CO_2$ concentration in the wake structure behind a Taylor bubble during long-term dissolution. The local concentration is reduced by the mole fraction shift from pure $CO_2$ to gas mixtures due to counter diffusion of $N_2$, $O_2$, which directly affects the mass transfer according to the two film theory. Additionally, a $CO_2$ comparison has been established between the global measured bubble volume reduction and the integral of the local dissolved $CO_2$ concentration field, which validates the high accuracy of both applied measuring techniques.

The wall dominant regime of Taylor bubbles enables reproducible and adjustable rise velocities, mass transfer coefficients and wake structures, by adjusting the channel diameter. The Taylor bubble experiment also enables detailed investigations of the influence of wake structures on chemical reactions and have therefore become one of the guidance experiments of the priority program DFG SPP1740[1] of the German Research Foundation.
The Taylor bubble experiment proved to be a valuable tool for detailed investigations of physi- and chemisorption in gas/liquid flows. The precise adjustment of the wake structure can be used to investigate reactive Taylor bubbles with the assumption of rotational symmetry to perform quantitative measurements with planar LIV techniques. Additionally, the constant rising velocity over a wide range of bubble sizes allows performing long-term investigations of the reaction products along the down-flow path with, e.g., Raman spectroscopy or other high-resolving measurement techniques with long integration times. The possibility to adjust the timescales of mixing to the timescales of chemical reactions is an advantage of this setup which allows investigating the effect of different wake structures on yield and selectivity of competitive parallel / consecutive reactions.

## 5.1 Outlook

Detailed global and local investigations on fluid dynamics and mass transfer processes regarding dissolving $CO_2$ bubbles in water filled vertical channels have been carried out during this work. Therefore, the variation of fluids for further experiments should be taken into account to extend established correlations to other fluid parameters. This research field would

[1] http://www.dfg-spp1740.de/

be restricted by the knowledge of physical fluid parameters, anyhow molecular modeling can help to get realistic values for diffusion coefficients and Henry constants, while focusing on free energy states of gas molecules being transferred into the liquid phase. Of course, qualitative experiments with less known fluid parameters can help to characterize transport processes with each other for building databases of mass transfer processes in various fluids and mixtures.

On the basis of this work, it was possible to transfer the Taylor bubble as a guidance tool to chemical systems within the priority program SPP1740 "Reactive bubble flows" of the German Research Foundation. Detailed analysis of reactive multiphase flows are still challenging due to the complex linked transport processes, but Taylor bubbles help to reduce the complexity, to lower the amount of required substances, to enlarge the experimental reproduction number and to perform high resolved global and local analysis of mass transfer processes and chemical reaction at the same time. Laser induced fluorescence can be used for detection of specific chemical substances, where their fluorescence properties can be calibrated with local concentrations of certain reactants or products, without any additional dye. Especially, competitive consecutive reactions are of high importance in chemical industry, where the adjustment of mixing timescales and their probable effect on yield and selectivity for competitive consecutive reactions with moderate reaction timescales, opens the field of process controlability. Those experiments would obviously depend on the chemical properties, their absorbance, emitting wavelength band and fluorescence intensity. Here, the coupling with measuring techniques like UV-vis or tomography techniques for a new chapter of research on industrial relevant reactive multiphase flows, would enable the quantification and validation of e.g. generated substances in the bubble wake.

In the future, reactive multiphase systems could be characterized in complexity reduced setups to get a first impression of the relation between reaction and mixing timescales and to characterize chemical systems within this benchmark experiments. The Taylor bubbles have a very high potential as a tool for the research as well as for the chemical industry. The requirement of small fluid volumes and amount of chemical substances make the Taylor bubble an ideal safe and reliable tool and minimizes the risk of hazardous accidents in chemical industry, for early investigations for presetting processes in the future. Qualitative description of unknown reactions as well as highly resolved quantitative investigations on mixing and reaction will be accessible for real and relevant processes of the chemical industry. Especially, for process safety and the sustainability of processes in gas/liquid contact apparatus, this guiding measuring system has a high potential in chemical industry in times of industry 4.0.

# References


[Abe08] Abe, S., Okawa, H., Hosokawa, S. and Tomiyama, A. *Dissolution of a Carbon Dioxide Bubble in a Vertical Pipe*. *J. Fluid Sci. Technol.*, 3(5):667–677, 2008. ISSN 1880-5558. doi:10.1299/jfst.3.667.

[Abo13] Aboulhasanzadeh, B., Hosoda, S., Tomiyama, A. and Tryggvason, G. *A validation of an embedded analytical description approach for the computations of high Schmidt number mass transfer from bubbles in liquids* . *Chem. Eng. Sci.*, 101(0):165–174, 2013. ISSN 00092509. doi:10.1016/j.ces.2013.06.020.

[Adr11] Adrian, R.J. *Particle image velocimetry*. Cambridge aerospace series. Cambridge Univ. Press, Cambridge u.a., 2011. ISBN 9780521440080 0521440084.

[Ala13] Aland, S., Lehrenfeld, C., Marschall, H., Meyer, C. and Weller, S. *Accuracy of two-phase flow simulations: The Taylor Flow benchmark*. *Pamm*, 13(1):595–598, 2013. ISSN 16177061. doi:10.1002/pamm.201310278.

[Alv05] Alves, S.S., Orvalho, S.P. and Vasconcelos, J.M.T. *Effect of bubble contamination on rise velocity and mass transfer*. *Chem. Eng. Sci.*, 60(1):1–9, 2005. ISSN 00092509. doi:10.1016/j.ces.2004.07.053.

[Aok13] Aoki, J., Hosoda, S., Hayashi, K., Hosokawa, S. and Tomiyama, A. *Mass transfer from single carbon dioxide bubbles in glycerol-water solution*. *Heat Mass Transf.*, 83(4):652–658, 2013. ISSN 00179310. doi:10.1016/j.ijheatmasstransfer.2014.12.062.

[Aok15] Aoki, J., Hayashi, K. and Tomiyama, A. *Mass transfer from single carbon dioxide bubbles in contaminated water in a vertical pipe*. *Int. J. Heat Mass Transf.*, 83:652–658, 2015. ISSN 00179310. doi:10.1016/j.ijheatmasstransfer.2014.12.062.

[Aoy16] Aoyama, S., Zun, I., Hayashi, K., Hosokawa, S. and Tomiyama, A. *Lift force acting on single bubbles in linear shear flow*. *Int. J. Multiph. Flow*, 1:0–5, 2016. ISSN 03019322. doi:10.1016/j.ijmultiphaseflow.2017.07.003.

[Ben18] Benders, S., Strassl, F., Fenger, B., Blümich, B., Herres-Pawlis, S. and Küppers, M. *Imaging of copper oxygenation reactions in a bubble flow*. *Magn. Reson. Chem.*, 56(9):826–830, 2018. ISSN 1097458X. doi:10.1002/mrc.4742.

[Bod14] Boden, S., Dos Santos Rolo, T., Baumbach, T. and Hampel, U. *Synchrotron radiation microtomography of Taylor bubbles in capillary two-phase flow*. *Exp. Fluids*, 55(7), 2014. ISSN 07234864. doi:10.1007/s00348-014-1768-7.

[Bor05] Bork, O., Schlueter, M. and Raebiger, N. *The Impact of Local Phenomena on Mass Transfer in Gas-Liquid Systems. Can. J. Chem. Eng.*, 83(4):658–666, 2005. ISSN 00084034. doi:10.1002/cjce.5450830406.

[Bra71] Brauer, H. *Stoffaustausch - einschließlich chemischer Reaktion.* Sauerländer AG, Aarau, 1971. ISBN 3794100085.

[Bre61] Bretherton, F.P. *The motion of long bubbles in tubes. J. Fluid Mech.*, 10:166–188, 1961. doi:10.1017/S0022112061000160.

[Bro65] Brown, R. *The Mechanics of Large Gas Bubbles in Tubes I. Bubble Velocities in Stagnant Liquids. Can. J. Chem. Eng.*, 43:217–223, 1965.

[Bug98] Bugg, J., Mack, K. and Rezkallah, K. *A numerical model of Taylor bubbles rising through stagnant liquids in vertical tubes. Int. J. Multiph. Flow*, 24(2):271–281, 1998. ISSN 03019322. doi:10.1016/S0301-9322(97)00047-5.

[Bug02] Bugg, J. and Saad, G. *The velocity field around a Taylor bubble rising in a stagnant viscous fluid: numerical and experimental results. International Journal of Multiphase Flow*, 28(5):791 – 803, 2002. ISSN 0301-9322. doi:https://doi.org/10.1016/S0301-9322(02)00002-2.

[But16] Butler, C., Cid, E. and Billet, A.M. *Modelling of mass transfer in Taylor flow: Investigation with the PLIF-I technique. Chemical Engineering Research and Design*, 115:292–302, 2016. ISSN 02638762. doi:10.1016/j.cherd.2016.09.001.

[But18] Butler, C., Lalanne, B., Sandmann, K., Cid, E. and Billet, A.M. *Mass transfer in Taylor flow: transfer rate modelling from measurements at the slug and film scale. International Journal of Multiphase Flow*, 0:1–17, 2018. ISSN 03019322. doi:10.1016/j.ijmultiphaseflow.2018.04.005.

[Cam88] Campos, J.B.L.M. and Carvalho, J.R.F.G.D. *An experimental study of the wake of gas slugs rising in liquids. Journal of Fluid Mechanics*, 196:27–37, 1988. doi:10.1017/S0022112088002599.

[Cli78] Clift, R., Grace, J.R., Weber, M. *Bubbles, Drops and Particles*, vol. 94. 1978. ISBN 012176950X. doi:10.1017/S0022112079221290.

[Dav50] Davies, R.M. and Taylor, G. *The mechanics of large bubbles rising through extended liquids and through liquids in tubes. Proc. R. Soc. London. Ser. A. Math. Phys. Sci.*, 200(1062):375–390, 1950. ISSN 2053-9169. doi:10.1098/rspa.1950.0023.

[Die13] Dietrich, N., Loubière, K., Jimenez, M., Hébrard, G. and Gourdon, C. *A new direct technique for visualizing and measuring gas – liquid mass transfer around bubbles moving in a straight millimetric square channel. Chem. Eng. Sci.*, 100:172–182, 2013. ISSN 0009-2509. doi:10.1016/j.ces.2013.03.041.

[Duk16] Dukhin, S.S., Lotfi, M., Kovalchuk, V.I., Bastani, D. and Miller, R. *Dynamics of rear stagnant cap formation at the surface of rising bubbles in surfactant solutions at large Reynolds and Marangoni numbers and for slow sorption kinetics. Colloids Surfaces A Physicochem. Eng. Asp.*, 492:127–137, 2016. ISSN 18734359. doi:10.1016/j.colsurfa.2015.12.028.

[Dum43] Dumitrescu, D.T. *Strömung an einer Luftblase im senkrechten Rohr. ZAMM - Zeitschrift für Angew. Math. und Mech.*, 23(3):139–149, 1943. ISSN 00442267. doi:10.1002/zamm.19430230303.

[Fil81] Filla, M. *Gas absorption from a slug held stationary in downflowing liquid . Chem. Eng. J.*, 22(3):213–220, 1981. ISSN 03009467. doi:10.1016/0300-9467(81)80016-0.

[Fra11] Francois, J., Dietrich, N., Guiraud, P. and Cockx, A. *Direct measurement of mass transfer around a single bubble by micro-PLIFI. Chem. Eng. Sci.*, 66(14):3328–3338, 2011. ISSN 00092509. doi:10.1016/j.ces.2011.01.049.

[Gib13] Gibson, A. *On the motion of long air-bubbles in a vertical tube. Philos. Mag. Ser. 6*, 26(156):952–966, 1913.

[Gri60] Griffith, R. *Mass transfer from drops and bubbles. Chemical Engineering Science*, 12(3):198–213, 1960. ISSN 00092509. doi:10.1016/0009-2509(60)85006-3.

[Hag16] Haghnegahdar, M., Boden, S. and Hampel, U. *Mass transfer measurement in a square milli-channel and comparison with results from a circular channel. Int. J. Heat Mass Transf.*, 101:251–260, 2016. ISSN 00179310. doi:10.1016/j.ijheatmasstransfer.2016.05.014.

[Hay11] Hayashi, K., Kurimoto, R. and Tomiyama, A. *Terminal velocity of a Taylor drop in a vertical pipe. Int. J. Multiph. Flow*, 37(3):241–251, 2011. ISSN 03019322. doi:10.1016/j.ijmultiphaseflow.2010.10.008.

[Hay12] Hayashi, K. and Tomiyama, A. *Effects of surfactant on terminal velocity of a Taylor bubble in a vertical pipe. Int. J. Multiph. Flow*, 39:78–87, 2012. ISSN 0301-9322. doi:10.1016/j.ijmultiphaseflow.2011.11.001.

[Hay14] Hayashi, K., Hosoda, S., Tryggvason, G. and Tomiyama, A. *Effects of shape oscillation on mass transfer from a Taylor bubble. Int. J. Multiph. Flow*, 58:236–245, 2014. ISSN 03019322. doi:10.1016/j.ijmultiphaseflow.2013.09.009.

[Hew69] Hewitt, G.F. and Roberts, D.N. *Studies of Two-Phase Flow Patterns by Simultaneous X-Ray and Flash Photography. Other Inf. UNCL. Orig. Receipt Date 31-DEC-69*, p. Medium: X; Size: Pages: 28, 1969.

[Hig35] Higbie, R. *The Rate of Absorption of a Pure Gas Into Still Liquid During Short Periods of Exposure*. 1935.

[Him64] Himmelblau, D.M. *Diffusion of dissolved gases in liquids. Chem. Rev.*, 64(5):527–550, 1964. ISSN 15206890. doi:10.1109/CLEO.2007.4452677.

[Hof06] Hoffmann, M., Schlüter, M. and Räbiger, N. *Experimental investigation of liquid–liquid mixing in T-shaped micro-mixers using -LIF and -PIV. Chem. Eng. Sci.*, 61(9):2968–2976, 2006. ISSN 00092509. doi:10.1016/j.ces.2005.11.029.

[Hor16] Hori, Y., Aoki, J., Hayashi, K., Hosokawa, S. and Tomiyama, A. *Mass transfer from single carbon dioxide bubbles in glycerol-water solution. Heat Mass Transf.*, 83(4):652–658, 2016. ISSN 00179310. doi:10.1016/j.ijheatmasstransfer.2014.12.062.

[Hor17] Hori, Y., Hayashi, K., Hosokawa, S. and Tomiyama, A. *Mass transfer from single carbon-dioxide bubbles in electrolyte aqueous solutions in vertical pipes. International Journal of Heat and Mass Transfer*, 115:663–671, 2017. ISSN 00179310. doi:10.1016/j.ijheatmasstransfer.2017.07.087.

[Hor19] Hori, Y., Hirota, Y., Hayashi, K., Hosokawa, S. and Tomiyama, A. *Combined effects of alcohol and electrolyte on mass transfer from single carbon-dioxide bubbles in vertical pipes. International Journal of Heat and Mass Transfer*, 136:521 – 530, 2019. ISSN 0017-9310. doi:https://doi.org/10.1016/j.ijheatmasstransfer.2019.02.094.

[Hor20] Hori, Y., Bothe, D., Hayashi, K., Hosokawa, S. and Tomiyama, A. *Mass transfer from single carbon-dioxide bubbles in surfactant-electrolyte mixed aqueous solutions in vertical pipes. International Journal of Multiphase Flow*, 124:103207, 2020. ISSN 0301-9322. doi:https://doi.org/10.1016/j.ijmultiphaseflow.2020.103207.

[Hos12] Hosokawa, S. and Tomiyama, A. *Spatial filter velocimetry based on time-series particle images. Exp. Fluids*, 52:1361–1372, 2012. ISSN 07234864. doi:10.1007/s00348-011-1259-z.

[Hos14a] Hosoda, S., Abe, S., Hosokawa, S. and Tomiyama, A. *Mass transfer from a bubble in a vertical pipe. Int. J. Heat Mass Transf.*, 69:215–222, 2014. ISSN 00179310. doi:10.1016/j.ijheatmasstransfer.2013.10.031.

[Hos14b] Hosokawa, S., Kitahata, K., Hayashi, K. and Tomiyama, A. *Evaluation of Interfacial Shear Stress of Single Drop using Spatiotemporal Filter Velocimetry High-speed. 17th International Symposium on Application of Laser Techniques to Fluid Mechanics, Lisbon, Portugal*, 2014.

[Hos15] Hosoda, S., Tryggvason, G., Hosokawa, S. and Tomiyama, A. *Dissolution of single carbon dioxide bubbles in a vertical pipe. J. Chem. Eng. Japan*, 48(6):418–426, 2015. doi:10.1252/jcej.14we241.

[Hos17] Hosokawa, S., Masukura, Y., Hayashi, K. and Tomiyama, A. *Experimental Evaluation of Marangoni Stress and Surfactant Concentration at Interface of Contaminated Single Spherical Drop Using Spatiotemporal Filter Velocimetry. Int. J. Multiph. Flow*, 2017. ISSN 03019322. doi:10.1016/j.ijmultiphaseflow.2017.08.007.

[Jia17] Jia, H.W. and Zhang, P. *Mass transfer of a rising spherical bubble in the contaminated solution with chemical reaction and volume change. International Journal of Heat and Mass Transfer*, 110:43–57, 2017. ISSN 00179310. doi:10.1016/j.ijheatmasstransfer.2017.02.095.

[Kas15] Kastens, S., Hosoda, S., Schlüter, M. and Tomiyama, A. *Mass Transfer from Single Taylor Bubbles in Mini Channels. Chem. Eng. Technol.*, (1):1–18, 2015. doi:10.1002/ceat.201500065.

[Kas17a] Kastens, S., Meyer, C., Hoffmann, M. and Schlüter, M. *Transport Processes at Fluidic Interfaces*. pp. 609–637, 2017. doi:10.1007/978-3-319-56602-3.

[Kas17b] Kastens, S., Timmermann, J., Strassl, F., Rampmaier, R., Hoffmann, A., Herres-Pawlis, S. and Schlüter, M. *Test System for the Investigation of Reactive Taylor Bubbles. Chem. Eng. Technol.*, 40(8), 2017. ISSN 15214125. doi:10.1002/ceat.201700047.

[Kin07] King, C., Walsh, E. and Grimes, R. *PIV measurements of flow within plugs in a microchannel. Microfluidics and Nanofluidics*, 3(4):463–472, 2007. ISSN 16134982. doi:10.1007/s10404-006-0139-y.

[Kra17] Krauβ, M. and Rzehak, R. *Reactive absorption of CO2 in NaOH: Detailed study of enhancement factor models. Chem. Eng. Sci.*, 166(March):193–209, 2017. ISSN 00092509. doi:10.1016/j.ces.2017.03.029.

[Kre05] Kreutzer, M.T., Kapteijn, F., Moulijn, J.A., Kleijn, C.R. and Heiszwolf, J.J. *Inertial and Interfacial Effects on Pressure Drop of Taylor Flow in Capillaries.* 51(9), 2005. doi:10.1002/aic.10495.

[Kri99] Krishna, R., Urseanu, M.I., Van Baten, J.M. and Ellenberger, J. *Wall effects on the rise of single gas bubbles in liquids. Int. Commun. Heat Mass Transf.*, 26(6):781–790, 1999. ISSN 07351933. doi:10.1016/S0735-1933(99)00066-4.

[Kur13] Kurimoto, R., Hayashi, K. and Tomiyama, A. *Terminal velocities of clean and fully-contaminated drops in vertical pipes. Int. J. Multiph. Flow*, 49:8–23, 2013. ISSN 03019322. doi:10.1016/j.ijmultiphaseflow.2012.08.001.

[Lew24] Lewis, W.K. and Whitman, W.G. *Principles of Gas Absorption. Ind. Eng. Chem.*, 16(12):1215–1220, 1924. ISSN 00197866. doi:10.1021/ie50180a002.

[LM65] La Mer, V.K. and Healy, T.W. *Evaporation of Water: Its Retardation by Monolayers. Science*, 148(3666):36–42, 1965. ISSN 00368075, 10959203.

[Luc11] Lucas, D. and Tomiyama, A. *On the role of the lateral lift force in poly-dispersed bubbly flows. Int. J. Multiph. Flow*, 37(9):1178–1190, 2011. ISSN 03019322. doi:10.1016/j.ijmultiphaseflow.2011.05.009.

[Mae75] Maeda, N. and Maeda, N. *Behavior of a single bubble in quiescent and flowing liquid inside a cylindrical tube. J. Nucl. Sci. Technol.*, 12(10):606–617, 1975. ISSN 00223131. doi:10.1080/18811248.1975.9733161.

[Mae15] Maeda, Y., Hosokawa, S., Baba, Y., Tomiyama, A. and Ito, Y. *Generation mechanism of micro-bubbles in a pressurized dissolution method. Exp. Therm. Fluid Sci.*, 60:201–207, 2015. ISSN 08941777. doi:10.1016/j.expthermflusci.2014.09.010.

[Mey14] Meyer, C., Hoffmann, M. and Schlüter, M. *Micro-PIV analysis of gas–liquid Taylor flow in a vertical oriented square shaped fluidic channel . International Journal of Multiphase Flow*, 67(0):140–148, 2014. ISSN 03019322. doi:10.1016/j.ijmultiphaseflow.2014.07.004.

[Mor16] Morgado, A., Miranda, J., Araújo, J. and Campos, J.B. *Review on vertical gas-liquid slug flow. Int. J. Multiph. Flow*, 85:348–368, 2016. ISSN 03019322. doi:10.1016/j.ijmultiphaseflow.2016.07.002.

[Nog06] Nogueira, S., Riethmuller, M.L., Campos, J.B. and Pinto, A.M. *Flow patterns in the wake of a Taylor bubble rising through vertical columns of stagnant and flowing Newtonian liquids: An experimental study*. *Chem. Eng. Sci.*, 61(22):7199–7212, 2006. ISSN 00092509. doi:10.1016/j.ces.2006.08.002.

[Olg13] Olgac, U. and Muradoglu, M. *Effects of surfactant on liquid film thickness in the Bretherton problem* . *Int. J. Multiph. Flow*, 48(0):58–70, 2013. ISSN 03019322. doi:10.1016/j.ijmultiphaseflow.2012.08.007.

[Pau18] Paul, M., Strassl, F., Hoffmann, A., Hoffmann, M., Schlüter, M. and Herres-Pawlis, S. *Reaction systems for bubbly flows*. *Eur. J. Inorg. Chem.*, pp. 2101–2124, 2018. ISSN 14341948. doi:10.1002/ejic.201800146.

[Pee53] Peebles, F. and Garber, H. *Studies on the motion of gas bubble in liquids*. *Chem. Engng Prog. Symp. Ser*, 49:88–97, 1953.

[Pes18] Pesci, C., Weiner, A., Marschall, H. and Bothe, D. *Computational analysis of single rising bubbles influenced by soluble surfactant*. *J. Fluid Mech.*, 856:709–763, 2018. ISSN 0022-1120. doi:10.1017/jfm.2018.723.

[Pin98] Pinto, A.M., Coelho Pinheiro, M.N. and Campos, J.B. *Coalescence of two gas slugs rising in a co-current flowing liquid in vertical tubes*. *Chem. Eng. Sci.*, 53(16):2973–2983, 1998. ISSN 00092509. doi:10.1016/S0009-2509(98)00121-3.

[Pop00] Pope, S.B. *Turbulent Flows*. Cambridge University Press, 2000. doi:10.1017/CBO9780511840531.

[Raf18] Raffel, M., Willert, C.E., Scarano, F., Kähler, C.J., Wereley, S.T. and Kompenhans, J. *Particle Image Velocimetry*. 1. Springer International Publishing, Cham, 2018. ISBN 978-3-319-68851-0. doi:10.1007/978-3-319-68852-7.

[Ros06] Rosso, D., Huo, D.L. and Stenstrom, M.K. *Effects of interfacial surfactant contamination on bubble gas transfer*. 61:5500–5514, 2006. doi:10.1016/j.ces.2006.04.018.

[Rüt18] Rüttinger, S., Spille, C., Hoffmann, M. and Schlüter, M. *Laser-Induced Fluorescence in Multiphase Systems*. *ChemBioEng Rev.*, 5(4):253–269, 2018. ISSN 21969744. doi:10.1002/cben.201800005.

[Sai15] Saito, T. and Toriu, M. *Effects of a bubble and the surrounding liquid motions on the instantaneous mass transfer across the gas-liquid interface*. *Chem. Eng. J.*, 265:164–175, 2015. ISSN 13858947. doi:10.1016/j.cej.2014.12.039.

[San15] Sander, R. *Compilation of Henry's law constants (version 4.0) for water as solvent*. *Atmos. Chem. Phys. Discuss.*, 15:4399 –4981, 2015. ISSN 1680-7324. doi:10.5194/acp-15-4399-2015.

[Sen93] Sena Esteves, M.T. and Guedes De Carvalho, J.R. *Liquid-side mass transfer coefficient for gas slugs rising in liquids*. *Chem. Eng. Sci.*, 48(20):3497–3506, 1993. ISSN 00092509. doi:10.1016/0009-2509(93)85005-A.

[Str16] Strassl, F., Timmermann, J., Schlüter, M. and Herres-Pawlis, S. *Kinetik der Sauerstoffaktivierung*. *GIT Labor-Fachzeitschrift*, (60):39–41, 2016.

[Str17] Strassl, F., Grimm-Lebsanft, B., Rukser, D., Biebl, F., Biednov, M., Brett, C., Timmermann, R., Metz, F., Hoffmann, A., Rübhausen, M. and Herres-Pawlis, S. *Oxygen Activation by Copper Complexes with an Aromatic Bis(guanidine) Ligand. Eur. J. Inorg. Chem.*, 2017(27):3350–3359, 2017. ISSN 10990682. doi: 10.1002/ejic.201700528.

[Tay61] Taylor, G.I. *Deposition of a viscous fluid on a plane surface. J. Fluid Mech.*, 9(March 2006):218, 1961. ISSN 0022-1120. doi:10.1017/S0022112060001055.

[Thu97] Thulasidas, T., Abraham, M. and Cerro, R. *Flow patterns in liquid slugs during bubble-train flow inside capillaries. Chemical Engineering Science*, 52(17):2947 – 2962, 1997. ISSN 0009-2509. doi:https://doi.org/10.1016/S0009-2509(97) 00114-0.

[Tim18] Timmermann, J. *Experimental analysis of fast reactions in gas-liquid flows*. Cuvilier Verlag Göttingen, 2018. ISBN 9783736988750. Doctoral thesis.

[Tom96] Tomiyama, A., Sou, A. and Sakaguchi, T. *Numerical Simulation of a Taylor Bubble in a Stagnant Liquid inside a Vertical Pipe. JSME International Journal Series B*, 39(3):517–524, 1996. ISSN 1340-8054. doi:10.1299/jsmeb.39.517.

[Tom02] Tomiyama, a. *Single bubbles in stagnant liquids and in linear shear flows. Work. Meas. Technol. (MTWS5), FZ*, pp. 3–19, 2002.

[Too58] Toor, H.L. ; Marchello, J. *Film-penetration -Model for Mass and Heat Transfer. A.I.Ch.E. Journal,*, 4(1):97–101, 1958. ISSN 1547-5905. doi:10.1002/aic. 690040118.

[Tri15] Tripathi, M.K., Sahu, K.C. and Govindarajan, R. *Dynamics of an initially spherical bubble rising in quiescent liquid. Nat. Commun.*, 6:6268, 2015. ISSN 2041-1723. doi:10.1038/ncomms7268.

[Tso07] Tsoligkas, A.N., Simmons, M.J. and Wood, J. *Influence of orientation upon the hydrodynamics of gas-liquid flow for square channels in monolith supports. Chemical Engineering Science*, 62(16):4365–4378, 2007. ISSN 00092509. doi: 10.1016/j.ces.2007.04.051.

[Tun76] Tung, K.W. and Parlange, J.Y. *Note on the motion of long bubbles in closed tubes-influence of surface tension. Acta Mechanica*, 24(3):313–317, 1976. ISSN 1619-6937. doi:10.1007/BF01190380.

[Uno56] Uno, S. and Kintner, R.C. *Effect of wall proximity on the rate of rise of single air bubbles in a quiescent liquid. AIChE Journal*, 2(3):420–425, 1956. ISSN 0001-1541. doi:10.1002/aic.690020323.

[Van02] Van Hout, R., Gulitski, A., Barnea, D. and Shemer, L. *Experimental investigation of the velocity field induced by a Taylor bubble rising in stagnant water. Int. J. Multiph. Flow*, 28(4):579–596, 2002. ISSN 03019322. doi: 10.1016/S0301-9322(01)00082-9.

[Van05] Vandu, C.O., Liu, H. and Krishna, R. *Mass transfer from Taylor bubbles rising in single capillaries. Chem. Eng. Sci.*, 60(22):6430–6437, 2005. ISSN 00092509. doi:10.1016/j.ces.2005.01.037.

[vB04] van Baten, J. and Krishna, R. *CFD simulations of mass transfer from Taylor bubbles rising in circular capillaries. Chem. Eng. Sci.*, 59(12):2535–2545, 2004. ISSN 00092509. doi:10.1016/j.ces.2004.03.010.

[VDI15] VDI. *VDI Heat Atlas*, vol. 1. 2015. ISBN 978-3-540-77876-9. doi:10.1017/CBO9781107415324.004.

[Via03] Viana, F., Pardo, R., Yánez, R., Trallero, J.L. and Jospeh, D.D. *Universal correlation for the rise velocity of long gas bubbles in round pipes. J. Fluid Mech.*, 494:S0022112003006165, 2003. ISSN 00221120. doi:10.1017/S0022112003006165.

[vK19] von Kameke, A., Kastens, S., Rüttinger, S., Herres-Pawlis, S. and Schlüter, M. *How coherent structures dominate the residence time in a bubble wake: An experimental example. Chemical Engineering Science*, 207:317–326, 2019. doi:10.1016/j.ces.2019.06.033.

[Wal69] Wallis, G. *One-dimensional two-phase flow*. McGraw-Hill, 1969.

[Whi62] White, E.T. and Beardmore, R.H. *The velocity of rise of single cylindrical air bubbles through liquids contained in vertical tubes. Chemical Engineering Science*, 17(5):351–361, 1962. ISSN 00092509. doi:10.1016/0009-2509(62)80036-0.

# Appendix A

# General parameters and dimensionless numbers of the channels

Table A.1 Summary of the characteristic values of Taylor bubbles in vertical channels $5.5 < D < 8$ mm

| $D$ | 5.5 mm | 6 mm | 7 mm | 8 mm |
|---|---|---|---|---|
| $Eo$ | 4.1 | 4.9 | 6.7 | 8.7 |
| $v_B$ | 2 mm s$^{-1}$ | 6 mm s$^{-1}$ | 22 mm s$^{-1}$ | 37 mm s$^{-1}$ |
| $Re$ | 11 | 36 | 153 | 303 |
| $k_L$ | $7.5 \cdot 10^{-5}$ m s$^{-1}$ | $1.2 \cdot 10^{-4}$ m s$^{-1}$ | $1.8 \cdot 10^{-4}$ m s$^{-1}$ | $2 \cdot 10^{-4}$ m s$^{-1}$ |
| $Sh$ | 216 | 392 | 675 | 836 |

# Appendix B

# Absorption and emission spectra of Fluorescein

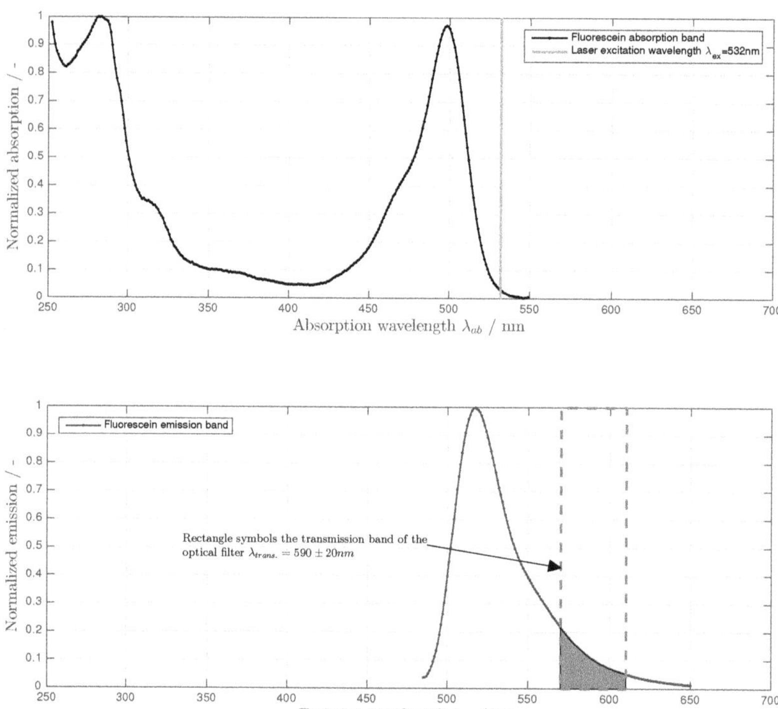


Fig. B.1 Absorption and emission spectra of fluorescein sodium salt according to fluorophores.org data bank. Excitation wavelength of the Laser and the filters transmission band protecting the camera are shown.

# Appendix C

# Settings for PIV and LIF equipment and post processing

Table C.1 Details of the PIV System, conditions of the experiments and parameters of post processing

| | |
|---|---|
| Type | 2D2C (two dimensional, two velocity components) |
| Laser | Quantronix Darwin Duo (Nd-YLF laser, 527 nm, up to 30 mJ) |
| Laser Sheet optics | ILA GmbH , sheet thickness approx. <1mm |
| Cameras | • PCO dimax HS2 - Nikkor 60 mm, protected by band-pass filter ($590 \pm 20nm$)<br>• PCO dimax HS4 - Nikkor 105 mm, protected by band-pass filter ($590 \pm 20nm$) |
| Seeding | • Microparticles GmbH, PS fluorescent particles, PS-FluoRed-Fi206, $d_{mean}$= 3.16 $\mu$m (or 1.61 $\mu$m)<br>• approximately 0.1 mL of (2.5 wt% ) suspension per 1 L of deionized water |
| number of image pairs | 300 - 1000 (decorrelated), depending on the flow field |
| Software | PIVview 2C 3.63 (PIVTec GmbH) |
| PIV processing parameters | • Multigrid Cross-Correlation<br>• 3 passes<br>• Final interrogation area size: 32 x 32 px to 24 x 24, depending on the<br>• Final vector spacing: 12 x 12 px<br>• Interrogation area windowing function: Welch window<br>• Image deformation: bicubic spline<br>• Image interpolation: logarithmic, bicubic spline<br>• Validation: floating median test<br>• Peak interpolation: 2 x 3 points, logarithmic |
| Spatial resolution | • Image scale: down to 43.2 $\mu$m/px<br>• Field of view: varying from 6 x 8 $mm^2$ to 6 x 20 $mm^2$ |
| Temporal resolution | • 600-1000 fps |

Table C.2 Details of the LIF System, conditions of the experiments and parameters of post processing

| Type | pLIF 2D (two dimensional planar ) |
|---|---|
| Laser | Quantronix Darwin Duo 527-100-M (Nd-YLF laser, 527 nm, up to 30 mJ at 1 kHz) |
| Laser Sheet optics | ILA GmbH , sheet thickness approx. <1mm |
| Camera Systems | • Highspeed cameras: PCO dimax HS2 & HS4<br>• Band-pass filter: Semrock 590/20 Brightline HC (590 $\pm$ 20 nm)<br>• Lenses: Zeiss Makro-Planar T* 50/2 ZF.2<br>• Long Range Distance Microscope: Infinity K2DistaMax -CF-3 (Field of View $\approx$ 3.5mm x 2.6 mm; working distance $\approx$ 9.5cm ) |
| Dye | • Fluorescein sodium salt, Sigma Aldrich (CAS Number 518-47-8)<br>• Dye concentration $C_{Dye}$=10 mmol m$^{-3}$<br>• low enough for a non measurable effect neither on the terminal rise velocity nor on overall mass transfer |
| Software | MATLAB 2018b Image Processing Toolbox, codes for calibration and application |
| LIF processing parameters | • laser sheet correction<br>• pH & $C_{CO_2}$ pixel wise calibration function |
| Spatial resolution | • Image scale $\approx$ 43.2 $\mu$m/px<br>• Field of view $\approx$ 44.2 x 44.2 mm$^2$ |
| Temporal resolution | • f=600-1000 fps |

# Appendix D

# Global mass transfer processes of Taylor bubbles in aqueous fluorescein solution

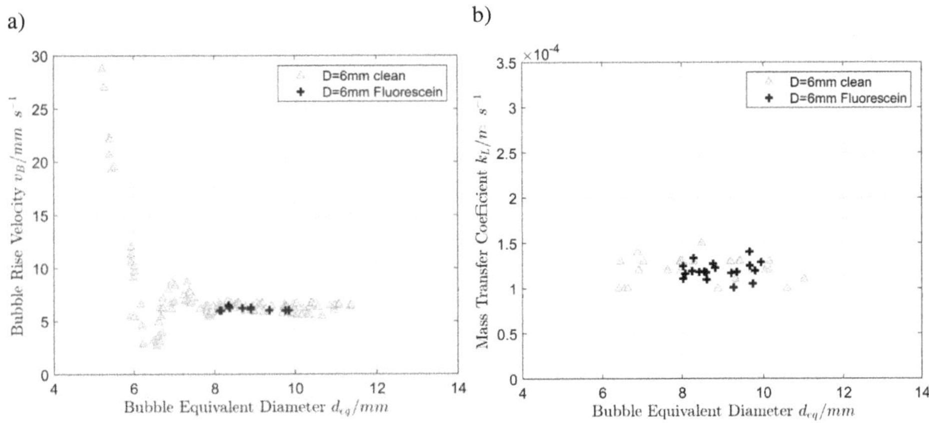


Fig. D.1 Non invasive concentration of the fluorescent dye fluorescein sodium salt ($C_{Dye} =$ 10 mmol m$^{-3}$) regarding the global measured transport processes $v_B$ and $k_L$. a) Rise Velocity $v_B$ of bubbles in the $D = 6$ mm channel in deionized water and fluorescein water solution; b) Comparison of mass transfer coefficient $k_L$ of Taylor bubbles in deionized water and fluorescein solution.

# Appendix E

# Local fluid dynamics at Taylor bubble front kept in CCF in circular channel $D = 6$ mm

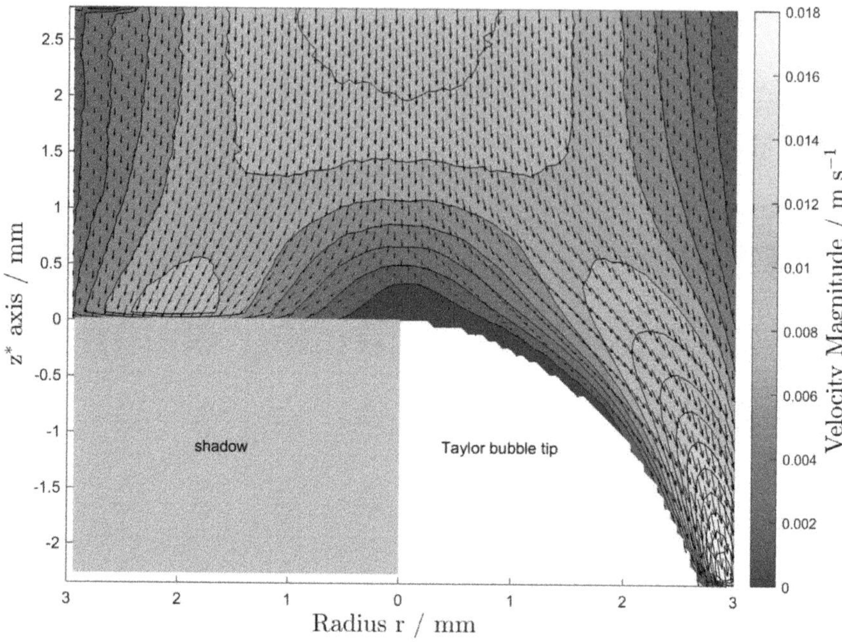


Fig. E.1 Local velocity field at a Taylor bubble front kept in counter current flow within the $D = 6$ mm circular channel.

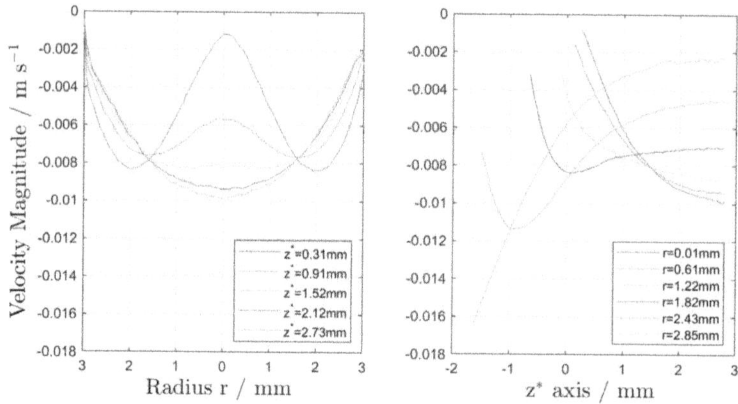


Fig. E.2 Local velocity profiles along constant height $z$ (left) or radius $r$ (right) at a Taylor bubble front kept in counter current flow within a $D = 6$ mm circular channel.

# Appendix F

# Local fluid dynamics at small bubble kept in CCF in square channel $D_h = 6$ mm

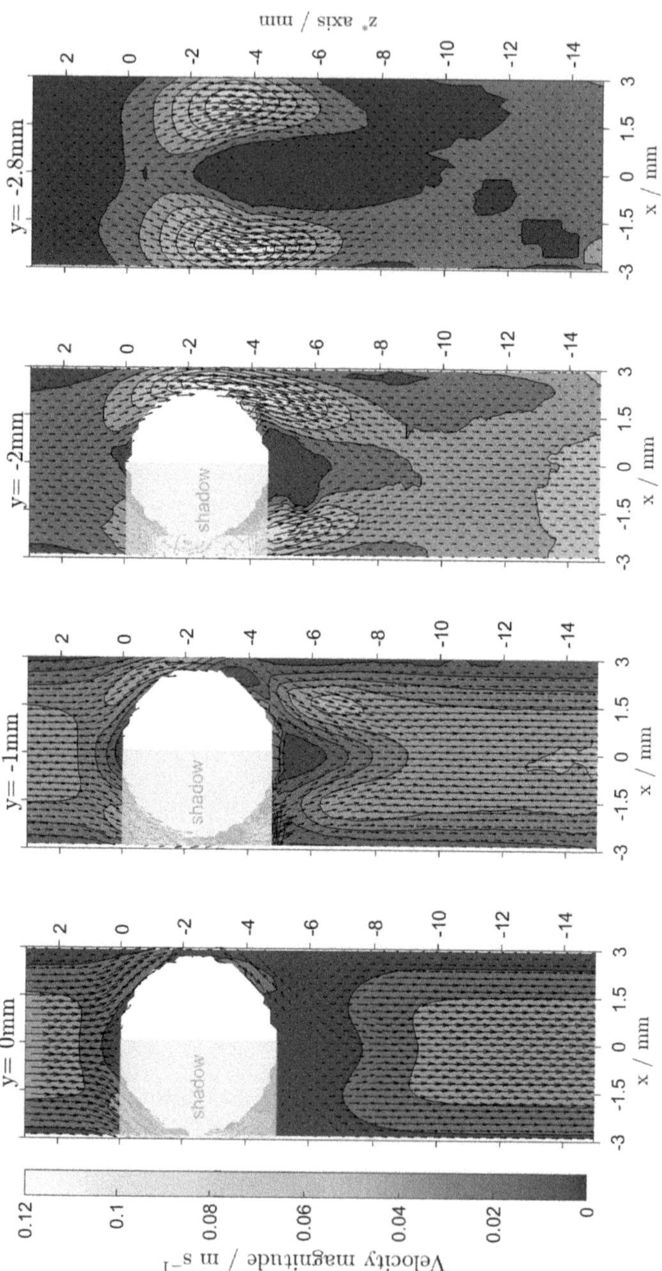


Fig. F.1 Overview of four 2D velocity fields in parallel planes at the identical small air bubble kept in CCF in $D_h = 6$ mm square channel

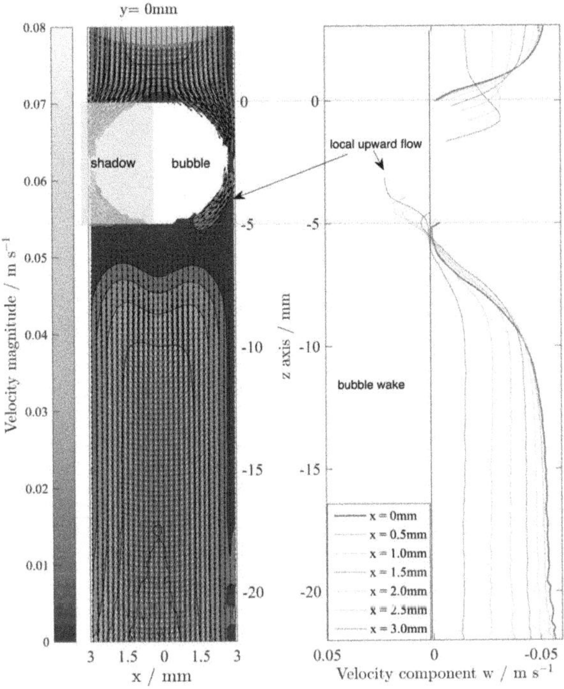


Fig. F.2 2D velocity fields at small air bubble kept in CCF in $D_h = 6$ mm square channel at $y = 0$ mm

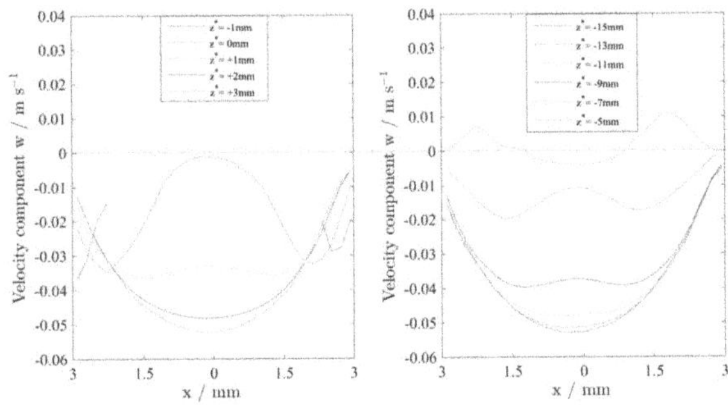


Fig. F.3 Velocity Profiles at small air bubble kept in CCF in $D_h = 6$ mm square channel at $y = 0$ mm

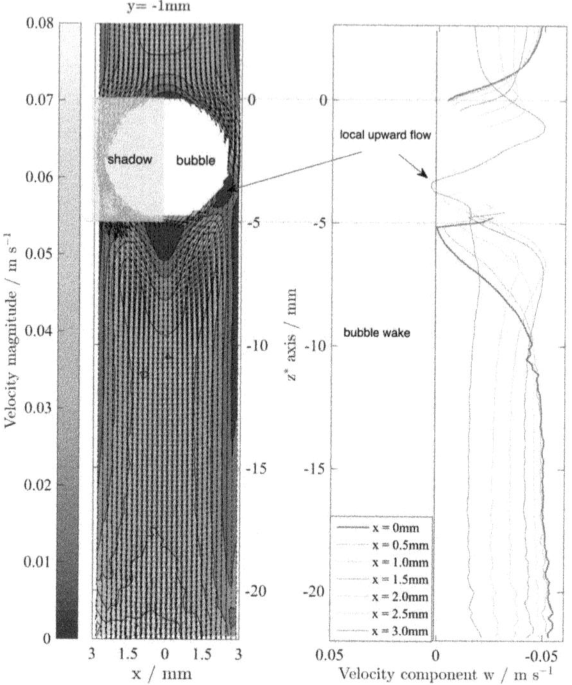


Fig. F.4 2D velocity fields at small air bubble kept in CCF in $D_h = 6$ mm square channel at $y = -1$ mm

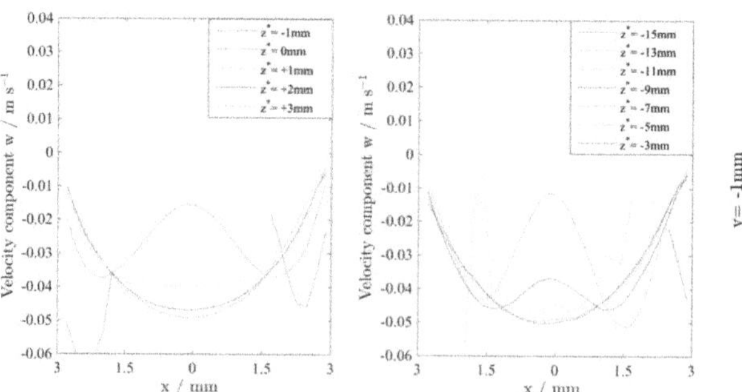


Fig. F.5 Velocity Profiles at small air bubble kept in CCF in $D_h = 6$ mm square channel at $y = -1$ mm

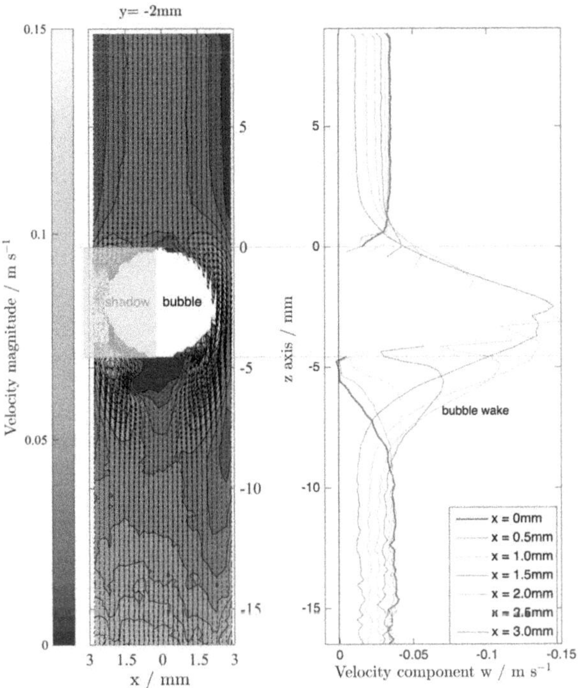


Fig. F.6 2D velocity fields at small air bubble kept in CCF in $D_h = 6$ mm square channel at $y = -2$ mm

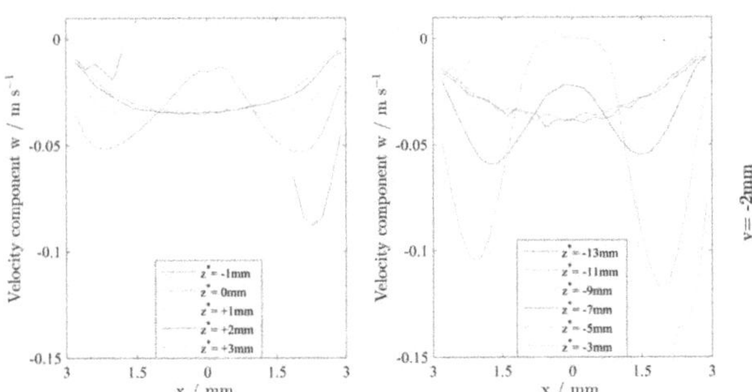


Fig. F.7 Velocity Profiles at small air bubble kept in CCF in $D_h = 6$ mm square channel at $y = -2$ mm

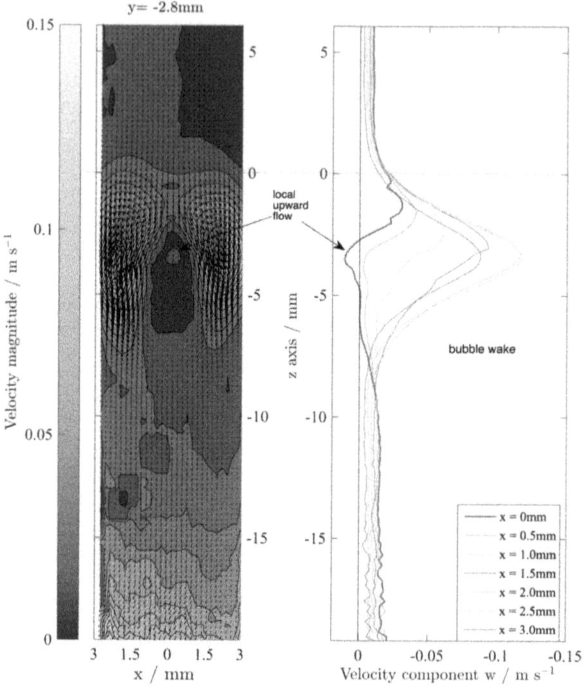


Fig. F.8 2D velocity fields at small air bubble kept in CCF in $D_h = 6$ mm square channel at $y = -2.8$ mm

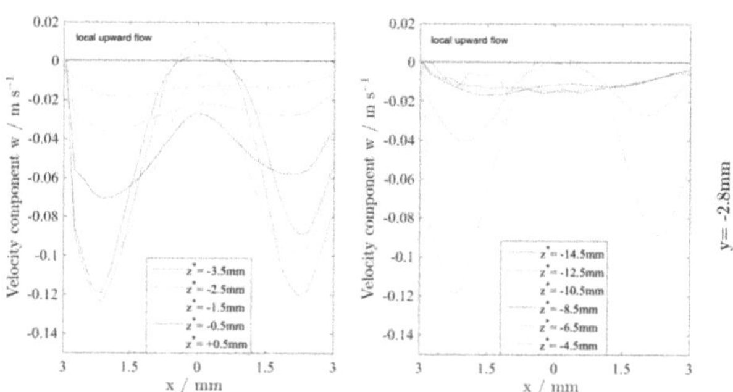


Fig. F.9 Velocity Profiles at small air bubble kept in CCF in $D_h = 6$ mm square channel at $y = -2.8$ mm

# Appendix G

# Taylor bubbles wakelength $l_{wake} = f(\nu)$ in square channel $D_h$ = 6 mm

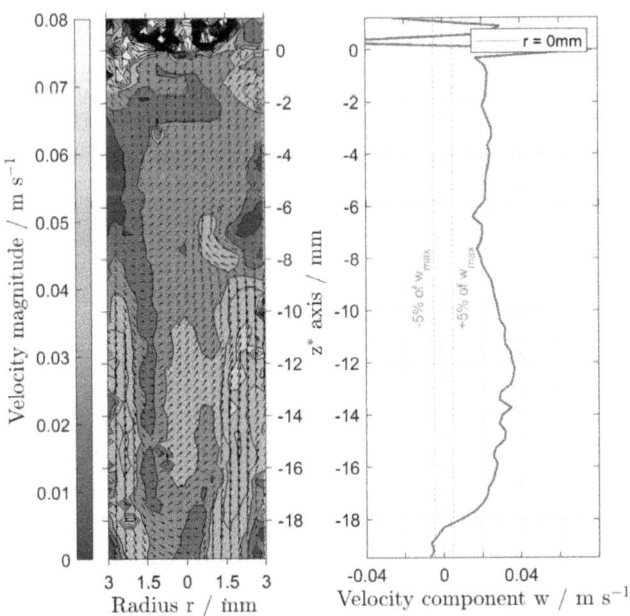


Fig. G.1 Wakelength measured in water. Left: Velocity field behind the Taylor bubble with a strong wake. Right: Velocity component $w$ in the channel center $r = 0$ mm.

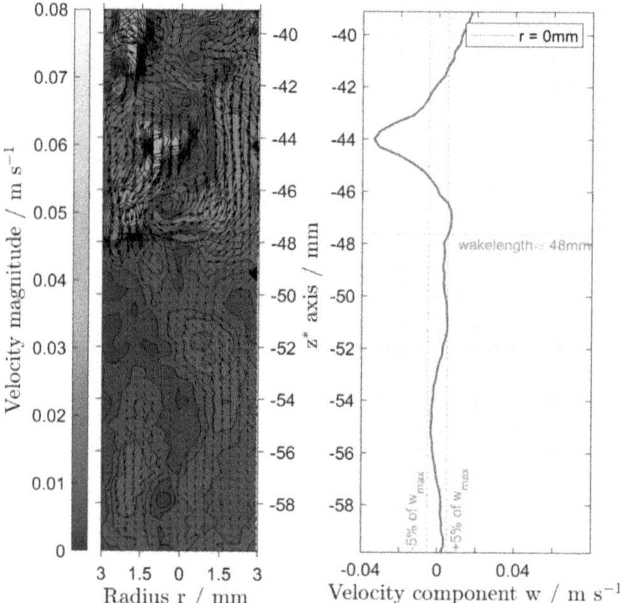


Fig. G.2 Wakelength measured in water. Left: Velocity field far behind the Taylor bubble. Right: Velocity component $w$ in the channel center $r = 0$ mm; estimated wakelength $l_{wake} \approx 48$ mm.

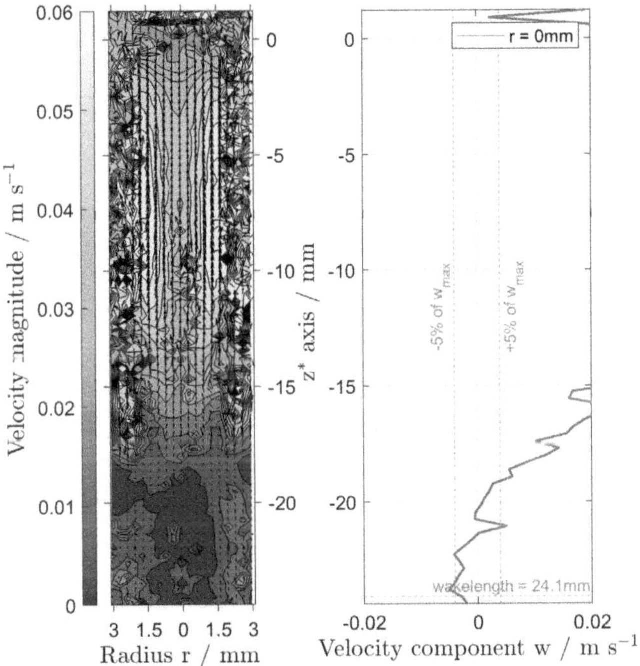


Fig. G.3 Wakelength measured in a glycerol 22 wt% water solution. Left: Velocity field far behind the Taylor bubble. Right: Velocity component $w$ in the channel center $r = 0$ mm; estimated wakelength $l_{wake} \approx 24$ mm.

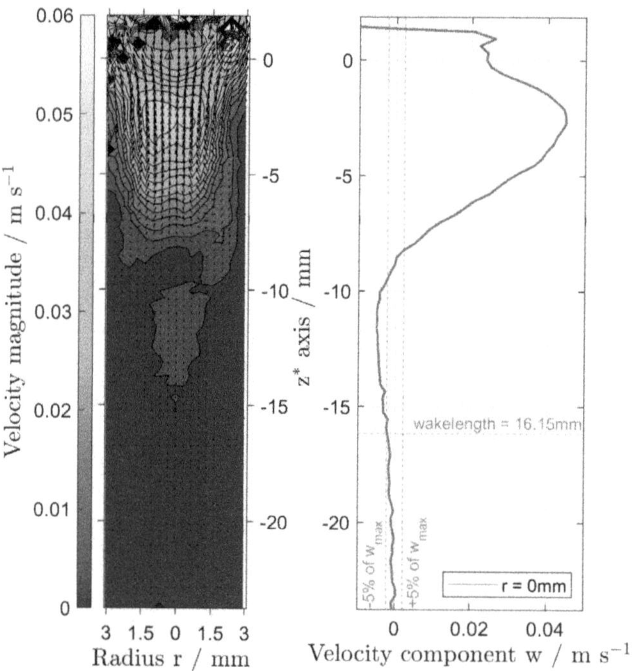


Fig. G.4 Wakelength measured in a glycerol 46 wt% water solution. Left: Velocity field behind the Taylor bubble. Right: Velocity component $w$ in the channel center $r = 0$ mm; estimated wakelength $l_{wake} \approx 16$ mm

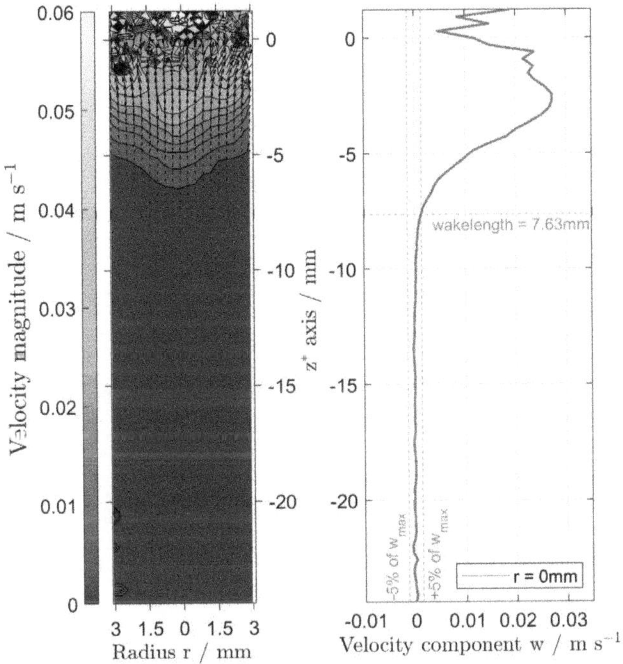


Fig. G.5 Wakelength measured in a glycerol 58 wt% water solution. Left: Velocity field behind the Taylor bubble. Right: Velocity component $w$ in the channel center $r = 0$ mm; estimated wakelength $l_{wake} \approx 7.6$ mm

# Appendix H

# Local fluid dynamics at clean and contaminated Taylor bubbles in circular channel $D = 6$ mm

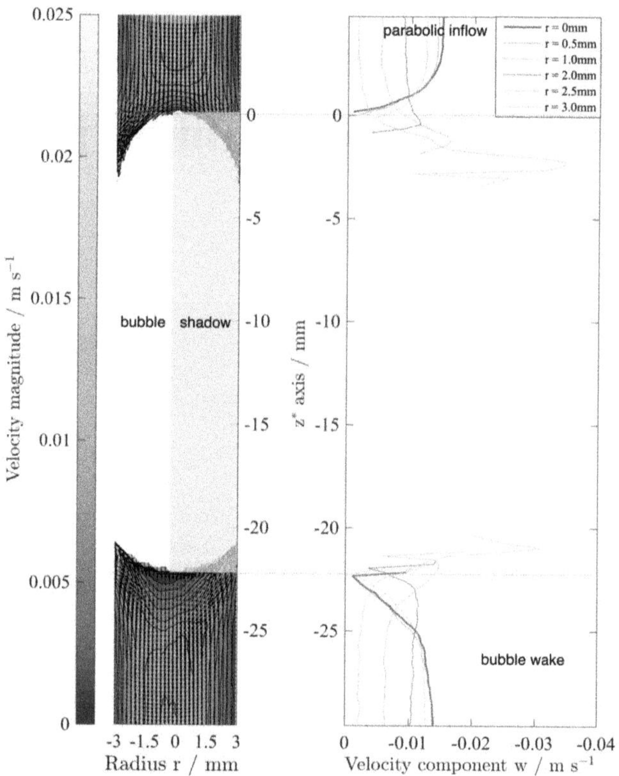


Fig. H.1 2D velocity fields in four parallel planes at the identical small air bubble in $D_h =$ 6 mm kept in CCF

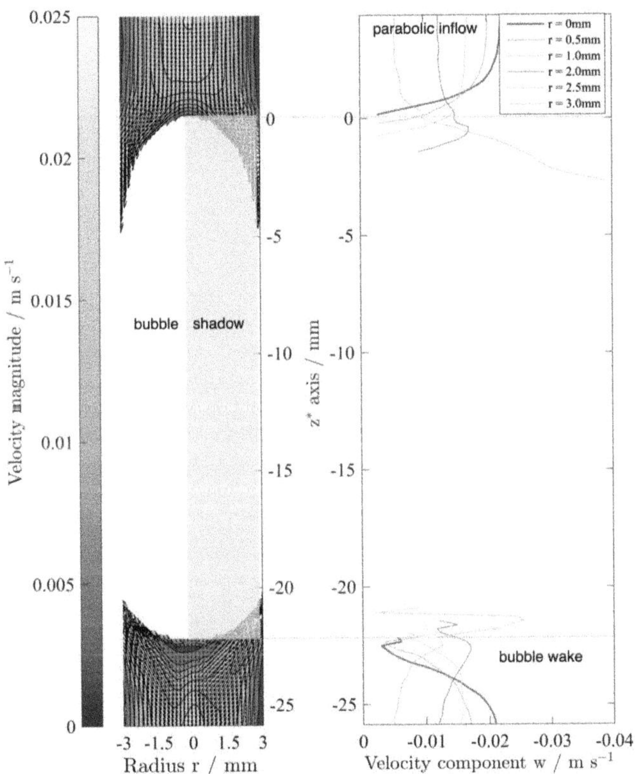


Fig. H.2 2D velocity fields in four parallel planes at the identical small air bubble in $D_h =$ 6 mm kept in CCF

# Appendix I

# Curriculum Vitae

| | |
|---|---|
| Name | Kastens |
| First Name | Sven |
| Date of Birth | 12.08.1985 |
| Place of Birth | Bremen |
| Country of Birth | Germany |

| | |
|---|---|
| 08/1992 - 07/1996 | Grundschule an der Glockenstraße (Primary School) in Bremen |
| 08/1996 - 07/2002 | Schulzentrum an der Parsevalstraße (Secondary School) in Bremen |
| 08/2002 - 07/2005 | Abitur at Gymnasiale Oberstufe des Schulzentrums an der Kurt-Schumacher-Allee (High School) in Bremen |
| 12/2005 - 09/2006 | Civil service at Deutsches Rotes Kreuz (German Red Cross), Kreisverband Hastedt in Bremen |
| 10/2006 - 09/2010 | Production Engineering at the University of Bremen, Major: Process Engineering, Degree: Bachelor of Science |
| 10/2010 - 07/2013 | Production Engineering at the University of Bremen, Major: Process Engineering, Degree: Master of Science |
| 08/2013 - 12/2018 | Doctoral Researcher and PhD student at Institute of Multiphase Flows at the Hamburg University of Technology |
| 10/2017 - 12/2018 | Group Leader 'Reactive Bubbly Flows' at Institute of Multiphase Flows at the Hamburg University of Technology |
| 07/2019 - today | Operations Manager, Sampling Department, Supply Chain Management, Aurubis AG, Hamburg |

www.ingramcontent.com/pod-product-compliance
Ingram Content Group UK Ltd.
Pitfield, Milton Keynes, MK11 3LW, UK
UKHW061826190726
13853UKWH00009B/2463

9 783736 974043